COEVOLUTION OF LIFE ON HOSTS

Coevolution

INTERSPECIFIC INTERACTIONS

A Series Edited by JOHN N. THOMPSON

# of Life on Hosts

INTEGRATING ECOLOGY AND HISTORY

DALE H. CLAYTON, SARAH E. BUSH,

AND KEVIN P. JOHNSON

THE UNIVERSITY OF CHICAGO PRESS   Chicago and London

Dale H. Clayton is professor of biology at the University of Utah. He is coeditor of *Host-Parasite Evolution: General Principles and Avian Models*, coauthor of *The Chewing Lice: World Checklist and Biological Overview*, and inventor of the LouseBuster™. **Sarah E. Bush** is an assistant professor of biology at the University of Utah. **Kevin P. Johnson** is a principal research scientist with the Illinois Natural History Survey at the University of Illinois at Urbana-Champaign. He is coauthor of *The Chewing Lice: World Checklist and Biological Overview*.

The University of Chicago Press, Chicago 60637
The University of Chicago Press, Ltd., London
© 2016 by The University of Chicago
All rights reserved. Published 2016.
Printed in the United States of America

24 23 22 21 20 19 18 17 16 15   1 2 3 4 5

ISBN-13: 978-0-226-30213-3 (cloth)
ISBN-13: 978-0-226-30227-0 (paper)
ISBN-13: 978-0-226-30230-0 (e-book)
DOI: 10.7208/chicago/9780226302300.001.0001

Library of Congress Cataloging-in-Publication Data
Clayton, Dale H., author.
    Coevolution of life on hosts : integrating ecology and history / Dale H. Clayton,
    Sarah E. Bush, and Kevin P. Johnson.
        pages cm — (Interspecific interactions)
        ISBN 978-0-226-30213-3 (cloth : alk. paper)—ISBN 978-0-226-30227-0 (pbk. : alk. paper)—
    ISBN 978-0-226-30230-0 (e-book) 1. Coevolution. 2. Lice. 3. Parasites. I. Bush, Sarah
    Elizabeth, author. II. Johnson, Kevin P., author. III. Title. IV. Series: Interspecific
    interactions.
        QH372.C53 2015
        576.8′7—dc23

                                                                                          2015011973

♾ This paper meets the requirements of ANSI/NISO Z39.48-1992 (Permanence of Paper).

We dedicate this book

to our students, the ones

who do the heavy lifting.

Probably our most important overall conclusion is

that the importance of reciprocal selective responses

between ecologically closely linked organisms has been

vastly underrated in considerations of the origins of

organic diversity.—Ehrlich and Raven 1964

CONTENTS

PREFACE

Coevolution is a hot topic. Over the past decade the number of papers on coevolution has increased by an order of magnitude. Research on coevolution is diverse, ranging from mathematical modeling, to experimental evolution studies, to cophylogenetic reconstruction. In this book we provide an introduction to coevolution in both microevolutionary (ecological) and macroevolutionary (historical) time. We emphasize the integration of cophylogenetic, comparative, and experimental approaches in testing coevolutionary hypotheses. The overriding question is "how do ecological interactions influence patterns of codiversification?"

The book is limited to the coevolution of interacting species. The term coevolution is sometimes used to describe antagonistic evolution between sexes *within* species, or the evolution of molecular interactions, such as those between RNA codons and anticodons. There is even a body of literature devoted to the coevolution of genes and culture. We address none of these. Rather, we restrict our attention to the coevolution of interactions between species, particularly those involving external parasites that live on hosts (rather than in them).

External parasites are a diverse assemblage of organisms, ranging from herbivorous insects on plants, to monogenean worms on fish, to mites on insects, to feather lice on birds. Ectoparasites are great models for studies of coevolution. They are relatively easy to observe, mark, and count. They can be eliminated to generate "clean" hosts for experiments, and they can be transferred between host individuals, populations, and species. Some forms of host defense, such as grooming behavior, can be experimentally manipulated to test the consequences of suboptimal defense on host fitness. Members of ectoparasite communities can be added to hosts in different combinations to study competitive interactions between parasites that share hosts.

Recent work in coevolutionary biology has been successful in demonstrating coadaptation between species in response to reciprocal selection. In contrast, few studies have tested the influence of coadaptation on diversification of interacting taxa. Short of inventing time travel, unraveling the role of different ecological factors in long-term macroevolutionary patterns is a serious challenge. However, we review attempts to do just this, some of which have met with moderate success. Many of our examples involve parasitic lice of birds and mammals, which represent unusually tractable systems for studies seeking to integrate ecology and history.

Lice are "permanent" parasites that pass all stages of their life cycle on the body of the host. They are closely associated with hosts in both micro- and macroevolutionary time. Some lice cospeciate with hosts, yielding congruent host-parasite phylogenies. Other lice do not cospeciate with hosts, and there is little or no concordance between their phylogenies. This variation in cophylogenetic history within a single order of insects is useful; it allows the exploration of ecological factors underlying very different macroevolutionary patterns without having to compare apples and oranges. By themselves, congruent phylogenies are not evidence for coadaptation; however, they are powerful frameworks for studies of coadaptation, as we illustrate in this book.

Although we frequently turn to lice, we also cover studies of plants, butterflies, moths, beetles, ants, fungi, wasps, bacteria, and other groups of coevolving organisms. In addition to coadaptive processes and cophylogenetic patterns, we consider factors that influence parasite dispersal and population structure, which are lynchpins between micro- and macroevolution. Rates of dispersal are central to evolutionary processes such as gene flow and random genetic drift. Dispersal arranges individuals and species into new interactions, altering selective regimes and the fitness landscape, as summarized by the geographic mosaic theory of coevolution. The *absence* of dispersal can reinforce interactions, even over long periods of time. Acting together, selection and dispersal provide much of the ecological context for host-parasite codiversification.

Studies of coevolution are often restricted to a single group of parasites, even though parasites seldom live in isolation. Indeed, many hosts support very diverse communities of parasites. Competitive interactions between these parasites can lead to character displacement and other forms of coadaptation between parasites. We show how interspecific competition in lice has fundamental effects on host use and diversification. We also show that broader community interactions, such as the phoretic dispersal of specialist parasites on generalist parasites, influence patterns of codiversification.

A major goal of this book is to illustrate how cophylogenetic, comparative, and experimental approaches can be integrated to demonstrate the influence of complex selective interactions on diversification. Reciprocal selection and coadaptation influence the codiversification of parasites and hosts, and the coadaptive radiation of parasites on hosts. Unfortunately, coevolutionary biology has a good deal of arcane terminology—enough to frighten even the most seasoned evolutionary biologist. We have therefore taken care to define the terminology we use, and to prepare figures illustrating relationships among different coevolutionary concepts and processes.

The book begins with an overview of the basic principles of coevolutionary biology, including both micro- and macroevolutionary approaches (chapter 1). In chapter 2 we provide an overview of the biology of lice, and in chapter 3 we survey the effects of lice on their hosts. Chapters 4–6 focus on coadaptive interactions in microevolutionary time. Chapter 4 reviews adaptations that mammals and birds have evolved to resist lice and other ectoparasites. Chapter 5 outlines counter-adaptations that lice have evolved to circumvent these host defenses. Chapter 6 reviews parasite competition, and the role of competition in the coadaptation of different species of lice that share hosts. Chapters 7 and 8 concern dispersal and population structure, initiating a transition from micro- to macroevolutionary time.

In chapters 9–12 we review patterns of codiversification between ectoparasites and their hosts, and the influence of different ecological parameters on these patterns. Chapter 9 explores the influence of cospeciation, host switching, and other processes on macroevolutionary patterns of association. In chapter 10 we consider the role of ecological factors, such as host resistance and parasite competition, in the dissimilar cophylogenetic histories of different groups of lice and their hosts. In chapter 11 we review the influence of coadaptation on the diversification of lice within single host species, and across multiple host species. Finally, in chapter 12 we explore the relationship of coadaptation to codiversification across a wider range of coevolving systems. We emphasize the need for more studies that integrate coadaptation and codiversification.

* * *

Much of the work described in this book was carried out by a series of outstanding students and postdocs with whom we have had the privilege of working. These talented people are the lifeblood of productive labs, including ours. We are extremely grateful to the following individuals, whose work is represented herein: Richard Adams, Heidi Campbell, Devin Drown, Bradley Goates, Chris Harbison, Dukgun Kim, Jennifer Koop, Patricia Lee, Jael Malenke, Brett Moyer, Vincent Smith, Wendy Smith, David Reed, Daniel Tompkins, Scott Villa, Jessica Waite, Bruno Walther, and Jason Weckstein. We are particularly grateful to Scott Villa for allowing us to use unpublished data from his ongoing thesis work.

We also wish to thank the following current members of our labs, whose comments improved the book: Julie Allen, Andrew Bartlow, Therese Cantanach, Emily Diblasi, Daniel Gustafsson, Sarah Knutie, Sabrina McNew, and Scott Villa. We are grateful to two anonymous reviewers, one of whom provided a 12-page single-spaced review of an early draft of the book. The comments of these reviewers, and those of series editor John Thompson, improved the book immeasurably. We are grateful to Mary Corrado for outstanding copyediting.

We thank the following colleagues for discussions related to various aspects of the book: Anurag Agrawal, Craig Benkman, Seth Bordenstein, Gerald Borgia, Butch Brodie III, Judie Bronstein, Stephen Cameron, Lissy Coley, Colin Dale, Jim Demastes, Hector Douglas, Lance Durden, Doug Futuyma, Franz Goller, Mark Hafner, Tom Kursar, Jessica Light, Jeb Owen, Rod Page, Ricardo Palma, Tom Parchman, Town Peterson, Roger Price, Heather Proctor, Ben Roberts, Mike Shapiro, Doug Schemske, Bruce Smith, Vincent Smith, Jason Weckstein, Kazunori Yoshizawa, and others who we have undoubtedly forgotten to mention (and to whom we apologize!). We thank Noah Whiteman and the Department of Ecology and Evolutionary Biology, University of Arizona, for sponsoring a mini-sabbatical for D. H. C. and S. E. B. in 2011, during which work on the book was initiated.

We are grateful to the following individuals for help acquiring figures and/or permission to use them: Jacquie Greff (Tonal Vision), Paul Weldon, and Koichiro Zamma. We thank Martin Turner for allowing us to use his spectacular scanning electron micrographs (SEMs) of lice. We are grateful to Karen Zundel for many forms of assistance, including tracking down obscure literature and help with the Literature Cited section. We thank Lynette and James White for providing an ideal writing retreat. We thank our families and friends for long-suffering patience over the years that we worked on the book.

We thank the staff of the University of Chicago Press, beginning with the late Susan Abrams, who frequently shared her love of books with D. H. C. when he was a graduate student at Chicago. We are extremely grateful to editor Christie Henry for her sensible advice and endless supply of patience and enthusiasm. We thank John Thompson for gentle guidance: "Do you really think you can get it done in a year?"

Finally, we are grateful to the National Science Foundation for uninterrupted funding of our work over many years, including an NSF-OPUS (Opportunities for Promoting Understanding through Synthesis) grant supporting the preparation of this volume.

I BACKGROUND

1   INTRODUCTION TO COEVOLUTION

Coevolution occurs on a very broad scale and
may comprise most of evolution.
—Van Valen 1983

L
ife began with simple interactions between simple molecules. Over vast periods of time, these interactions increased in complexity, yielding macromolecules capable of
transmitting heritable information. More intricate levels of
biological organization followed, leading to the evolution
of cells, tissues, and whole organisms. The main mechanism underlying
these evolutionary changes involved random genetic mutations that coded
for phenotypic changes, most of which had negative or neutral effects on
fitness. However, a tiny fraction of these changes were adaptive, in that they
improved the ability of organisms to deal with environmental pressures.
Individuals with these adaptations left more offspring and thus greater
genetic representation. In this way, the adaptive changes quickly became
more common. This is the process Darwin (1859) called *evolution by natural
selection*.

Populations of individuals evolve in heterogeneous natural environments that exert selection for a multitude of solutions to a maze of challenges. Environmental heterogeneity subdivides populations, further increasing diversity through restricted gene flow and random genetic drift.
Over time, evolution by natural selection has generated a dizzying array of
life forms of mind-boggling complexity and beauty. It is these species that
make up the immense diversity of life on earth.

Species do not live in isolation; they live and interact with other species.
The history of life has been profoundly influenced by these interactions.
Interactions between species are so pervasive that they are a central feature
of the environments of nearly all organisms. Many interactions are selective
in nature, with one species influencing the fitness of other species. These
selective effects can lead to evolutionary changes in the affected species. In
some cases, the selective effects are reciprocal. When reciprocal selective
effects lead to evolutionary changes in two or more interacting species, the
process is known as *coadaptation*, or *coevolution* in the strict sense (box 1.1).

Darwin (1859) did not use the term coevolution, but he did refer to "coadaptations of organic beings to one another" (Thompson 1982, 1989). A century later, in a paper published in the journal *Evolution*, Mode (1958) used

BOX 1.1. Co-confusing concepts

Adaptation Microevolution (genetic change) that increases fitness (survival or reproductive success) in response to natural selection.

Adaptive codiversification Correlated diversification of interacting lineages in response to unidirectional selection on one of the lineages.

Coadaptation Microevolution of two or more interacting species in response to reciprocal selection between them (Janzen 1980).

Codiversification Correlated diversification of interacting lineages (Janz 2011; Althoff et al. 2014).

Coevolution In the strict sense, equivalent to coadaptation. Also used more broadly for the evolution of one species in response to its interaction with another species, or for the history of joint divergence of ecologically associated species (Futuyma 2013).

Coadaptive codiversification Correlated diversification of interacting lineages in response to reciprocal selection between them.

Coadaptive diversification Diversification of one lineage in response to reciprocal selection between interacting lineages; similar to "coevolutionary diversification" (Althoff et al. 2014) or "diversifying coevolution" (Thompson 2005).

Cophylogenetics Comparing phylogenies of interacting lineages (de Vienne et al. 2013).

Cospeciation Concomitant speciation of ecologically interacting lineages (Brooks 1979).

Diversification An evolutionary increase in the number of species in a clade, usually accompanied by divergence in phenotypic characters (Futuyma 2013).

---

coevolution in reference to reciprocal microevolutionary changes between agricultural crops and fungal pathogens in ecological time. Soon thereafter, in another paper published in *Evolution*, Ehrlich and Raven (1964) used coevolution to describe reciprocal adaptive radiations between butterflies and their food plants (plate 1a). Although both papers were concerned with selective interactions between species, they took very different approaches, one focusing on microevolutionary time, and the other on macroevolutionary time.

The divide between micro- and macroevolutionary approaches to coevolution continues. Microevolutionary studies test one or more assumptions underlying the process of coadaptation, such as reciprocal selection, heritability of traits, or the extent of evolutionary responses to selection. By contrast, macroevolutionary studies compare the phylogenetic histories of interacting groups, such as parasitic lice and their hosts (plate 2), in order to understand patterns of codiversification. Microevolutionary studies thus tend to focus on evolutionary ecological processes, while macroevolutionary studies tend to focus on historical patterns. Regardless of the time scale, both approaches test the null hypothesis that coevolution has *not* taken place, until proven otherwise.

Coevolution has been pervasive in the history of life. Complex organisms depend on interspecific interactions to survive and reproduce (Thompson 2009). Eukaryotic organelles such as mitochondria and chloroplasts are descendants of ancient, independent lineages that evolved mutualistic relationships with primitive host cells, and ultimately became part of those cells. In short, coevolutionary dynamics have played a major role in generating the diverse catalog of ecological interactions that make up Darwin's (1859) "tangled bank." These relationships include a large variety of antagonistic interactions, like those between hosts and parasites, predators and prey, and competing species. Coevolved relationships also include a diversity of mutually beneficial interactions, like those between pollinators and flowering plants, dinoflagellates and corals, endosymbiotic bacteria and insects, and the fungi and algae that make up lichens.

Coadaptation

Despite this incredible diversity of coadapted systems, effects between species boil down to one of two possibilities: Species A can have a positive effect on Species B, or it can have a negative effect. The reciprocal also holds: Species B can have a positive effect on Species A, or it can have a negative effect. Interactions between species thus generate four broad outcomes: mutualism, parasitism, predation, or competition (fig. 1.1). *Mutualism* is an interaction in which both parties benefit. The remaining three combinations are antagonistic interactions. In two of these cases, *predation* and *parasitism*, one party benefits, while the other is harmed. Predatory interactions are brief, lasting minutes or hours. Parasitic interactions are more durable, lasting days, weeks, or even years (Combes 2001). In the fourth type of interaction—*competition*—both parties are harmed. Interactions in which one species benefits, but the other is unaffected, are called *commensalism*. Examples of commensalism include bryophytes in the canopies of tropical trees, or remora fish dining on the leftovers of foraging sharks. Commensalistic interactions are not examples of coadaptation because they are unidirectional, by nature.

The relative fitness consequences of interspecific interactions can vary, depending on broader environmental contexts. For example, the effect of brood parasitic cuckoos on carrion crows fluctuates from parasitic to mu-

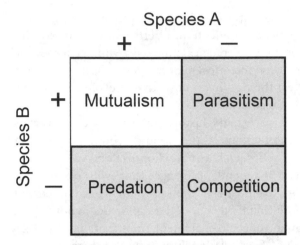

	Species A	
	+	**−**
Species B **+**	Mutualism	Parasitism
Species B **−**	Predation	Competition

FIGURE 1.1. Interspecific effects on fitness. In the case of predation, Species A has a negative effect on B, which is usually smaller in body size than A. In the case of parasitism, Species B has a negative effect on A, which is usually larger in body size than B. Shaded squares are forms of antagonistic interaction. See text for details.

tualistic, depending on the amount of crow nest predation by mammals and birds in a given year. When predators are abundant, the cost of rearing cuckoo nestlings is offset by the fact that cuckoo nestlings produce a repellent secretion that deters nest predators, thus increasing the fledging success of carrion crows (Canestrari et al. 2014). Hence, interspecific interactions are not static; they can transition between categories (Bronstein 1994).

It is important to bear in mind that, for fitness effects to generate adaptation, those effects must be *selective*. An effect of one species on another species is selective only if it generates a correlation between phenotypic variation and fitness. If the foreheads of hummingbird pollinators brush up mainly against stamens of flowers with deep corollas, then only flowers with deep corollas will get pollinated. In contrast, if hummingbirds brush up against flower stamens at random with respect to corolla depth, then even though these pollinators have a clear effect on plant fitness, they will not select for a change in corolla depth.

One reason why coadaptation is fascinating is that it has the power to explain remarkable examples of matching traits in nature. One of the most famous cases of matching traits is the Malagasy star orchid (*Angraecum sesquipedale*) and its pollinator, the Morgan hawkmoth (*Xanthopan morgani*) (fig. 1.2). In his book on orchid pollination, Darwin (1862) wrote about this orchid and its 30-cm-long nectar spur. After experimentally removing pollinia from specimens of *A. sesquipedale* flowers in different ways, Darwin predicted that the pollinator had to be an undescribed hawkmoth with a proboscis "capable of extension to a length of between 10 and 11 inches [25 centimeters]!" Darwin's prediction was ridiculed by some of his col-

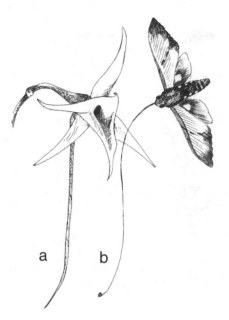

FIGURE 1.2. (a) The Malagasy star orchid (*Angraecum sesquipedale*) has a nectar spur that is about 30 cm long; (b) the orchid's pollinator, Morgan's hawkmoth (*Xanthopan morgani*), has a proboscis that is also about 30 cm long (drawn by S.E.B.).

a   b

leagues (van der Cingel 2001; Rodríguez-Gironés and Santamaría 2007). However, 21 years after his death, Darwin was vindicated when Rothschild and Jordan (1903) described *Xanthopan morgani praedicta*, a Malagasy hawkmoth with a four-inch-long (10 cm) body and 12-inch-long (30 cm) proboscis (fig. 1.2b). One cannot help but marvel at the scale, which is equivalent to a six-foot-tall man with an 18-foot-long tongue.

Matching traits need not be morphological; they can involve physiological, behavioral, and other traits. For example, parsnip webworm moth larvae (*Depressaria pastinacella*) feed exclusively on wild parsnips (*Pastinaca sativa*) over much of their range (plate 1b) (Zangerl and Berenbaum 2003). Wild parsnip produces furanocoumarins, which are highly toxic defensive compounds (Berenbaum 1991). The webworms detoxify these compounds using P450 proteins. The activity of the proteins is variable, and a target of selection by host plants (Berenbaum and Zangerl 1992). Across most populations, there is a close match between the mean furanocoumarin concentration in wild parsnip seeds and the rate of detoxification of furanocoumarins by webworms (Zangerl and Berenbaum 2003). This system is a classic example of chemical coadaptation.

Despite their intuitive appeal, it is important to keep in mind that matching traits, by themselves, are not evidence of coadaptation. Some matching traits have evolved due to historical interactions with entirely different species. Traits may have evolved in allopatric species that come into sym-

FIGURE 1.3. Cleaning mutualism in which coati mundis (*Nasua narica*) remove ticks from the ear canals of tapirs (*Tapirus bairdii*); see text (drawn by S.E.B. from photographs by D.H.C.).

patry at a later date and happen to fit one another, a phenomenon known as *ecological fitting* (Janzen 1985a). In an important paper with a citation count that exceeds its word count, Janzen (1980) "call[ed] for more careful attention to the use of 'coevolution' as a word and concept." He argued that casual application of the concept makes it synonymous with the general notion of interspecific interactions.

A tongue in cheek (or snout in ear) example of matching traits is the close fit between the snouts of coati mundis (*Nasua narica*) and the ears of tapirs (*Tapirus bairdii*). Coatis service tapirs by removing ticks from their ears (fig. 1.3). However, nobody in his right mind would argue that coati snouts and tapir ears are coadaptations. Indeed, the case in figure 1.3 turned out to be a short-lived interaction between a small number of tapirs and coatis looking for handouts near the dining hall on Barro Colorado Island, Panama, in the 1970s. Once visitors to the island stopped feeding the tapirs, the cleaning mutualism between coatis and tapirs disappeared (McClearn 1992).

In short, matching traits have not coevolved unless they have undergone coadaptation in response to reciprocal selection between the species possessing them. Note also that it is entirely possible for different categories of traits to coevolve—for example, a morphological feature in one species can coevolve with a physiological feature in another species. All that matters is that the trait in one species evolves in response to selection imposed by a second species. The evolutionary change then generates reciprocal selection and evolution in a trait of the second species. This criterion of reciprocity is widely accepted by coevolutionary biologists studying co-adaptation (Dawkins and Krebs 1979; Kiester et al. 1984; Thompson 1989, 2005; Wade 2007; Nuismer et al. 2010). The key concept, and one that makes coevolution unique, is that coadaptation leads to the partial coordination of non-mixing gene pools over evolutionary time.

The phenotypic interface of coadaptation

Rigorous tests of coadaptation attempt to isolate and quantify reciprocal selective effects at the *phenotypic interface* (fig. 1.4), which is influenced by a range of morphological, behavioral, physiological, and other performance traits. These traits can often be measured directly. The effect of one species on the performance of another species, and vice versa, governs the fitness of each species, which in turn governs how fitness correlates with phenotypic variation in each species. In other words, reciprocal effects on performance dictate reciprocal selection on corresponding traits between coadapting species. The performance component of the phenotypic interface, and the ability to quantify it, are essential ingredients in tests of coadaptation. No matter how perfectly traits seem to match, it is hard to know whether they are coevolving without knowing something about the reciprocal selective effects on those traits. A typical method of isolating and quantifying recip-rocal selective effects is to experimentally manipulate a trait in one species, then quantify the fitness consequences in the other species (Clayton et al. 1999; Nash 2008).

Measuring selection is also important in cases where matching traits of interacting species covary across geographic regions. Anderson and Johnson (2008) documented striking correlations among different locali-ties between *flower* corolla length (*Zaluzianskya microsiphon*) and proboscis length in a fly pollinator (*Prosoeca ganglbaueri*). Such patterns are intrigu-ing, and may well represent different coadaptive end points. However, theo-retical models suggest that geographic correlations can also arise due to selection on the members of one group to match a preexisting distribution of phenotypes in the other group (Nuismer et al. 2010). Geographic correla-tions can also be generated by independent bouts of stabilizing selection if the optimal phenotype for each species happens to correlate with some third abiotic or biotic factor.

Reciprocal selection cannot result in coadaptive evolution unless the

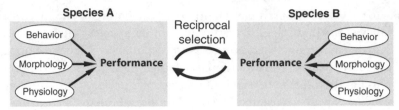

FIGURE 1.4. Reciprocal selection at the phenotypic interface of coevolution.
After Brodie and Ridenhour (2003).

phenotypic targets of selection have heritable genetic components. Quantitative genetic theory (Falconer 1981) tells us that the evolutionary response (R) of a trait is the product of the intensity of selection on that trait (s) and its narrow sense heritability (h^2):

$$R = h^2 s$$

Strong selection can exhaust genetic variation, meaning that traits under selection may have low heritabilities (Mousseau and Roff 1987). In short, reciprocal selection on the traits of interacting species is a necessary but insufficient condition for predicting coadaptive evolutionary responses. One also needs information concerning the genetic basis of the traits.

Other important genetic factors include whether the genome is haploid or diploid, the number of alleles influencing a given trait, and the possible interactions between those alleles, such as dominance, pleiotropy, or epistasis (Woolhouse et al. 2002). Coevolutionary dynamics may further be influenced by 1) the type of reproduction (sexual or asexual), 2) whether mating is random or assortative, 3) whether generations are discreet or continuous, and 4) the relative generation times of interacting species. Random genetic drift can also play an important role in the coevolutionary dynamics of small populations (Kiester et al. 1984).

Yet another important consideration in predicting coadaptive outcomes is that the evolutionary response to selection depends on environmental context. Fitness cannot be described as a function of phenotypes alone; the "value" of the environment encountered by an individual must also be considered (Brodie and Ridenhour 2003). A genotype (G) with high fitness in one environment (E) may not have high fitness in another environment. Thus, genotype-by-environment (G×E) interactions have a fundamental influence on evolutionary responses to selection.

From a coevolutionary viewpoint, a key component of the environment for one species is the other species with which it interacts. This other species has an evolutionary trajectory of its own and, hence, it effectively has the ability to "evolve back." The evolutionary response to selection will be influenced by variation in the selective agent (the other species), as well as variation in the abiotic and other biotic components of the environment. Thus, the outcome of selection on a given species depends partly on the standing genetic variation in populations of the other species. As noted by Wade (2007), "The reciprocal co-evolution of one species in response to the genetic context of another is integral to the process of co-adaptation between species, and is believed to contribute to the functional integration of ecological communities and maintenance of biodiversity." Coadaptive dynamics are complicated, and intriguing, because they depend on genotype-by-genotype-by-environment (G×G×E) interactions (Piculell et al. 2008; Sadd 2011).

The geographic mosaic theory of coevolution

G×G×E interactions are central to the geographic mosaic theory of co-evolution (GMTC) (Thompson 1994, 2005). Coadaptation is inherently a geographically structured process. Species are made up of populations that vary genetically. These populations are distributed among environments that vary because of differences in abiotic factors as well as local communities that vary in species composition (Hoeksema 2012). The GMTC has three distinct components governing coadaptation: 1) coevolutionary hotspots and coldspots, 2) selection mosaics, and 3) trait-remixing (fig. 1.5). Hot spots are areas in which reciprocal selection is strong, while cold spots are areas in which reciprocal selection is absent. The most extreme example of a cold spot is when one of the two species is not even present. In the case of host-parasite systems, this would be true if a local host population happened to be parasite free.

The second component, often misunderstood, is the concept of a selection mosaic, in which the type of selection varies among populations of a given species (fig. 1.5). Two hotspots can have equally intense reciprocal selection acting between species, yet selection may favor an increase in a trait in one location, while favoring a decrease in the same trait in another location. The important point about selection mosaics is that coadaptive outcomes can vary among populations, not just because the intensity of reciprocal selection varies, but also because the *type* of reciprocal selection varies. A good example of this is conditional mutualism, in which interactions between two species can be mutualistic, commensalistic, or parasitic depending on the composition of other members of the local community (Cushman and Whitham 1989; Thompson and Cunningham 2002).

The third component of the GMTC is trait-remixing, in which coevolved traits move between populations by gene flow, while genetic variation is maintained by differential mutation, random genetic drift, and meta-population dynamics. Hybridization, polyploidization, and other genomic processes also influence trait remixing (Thompson 2005). Continuous

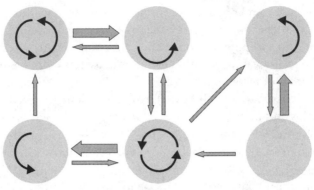

FIGURE 1.5. Geographic Mosaic Theory of Coevolution (GMTC) between two species. Circles represent local communities, with selective effects shown by black arrows (two circles show reciprocal selection, three circles show unidirectional selection, and one circle shows no selection). Differences in arrow angle represent differences in how selection acts on species in different communities (e.g., due to differences in the strength of selection). Arrows between circles represent gene flow between communities, with thicker arrows showing more gene flow. Coevolutionary hotspots (circles with reciprocal selection) occur within the broader matrix of coevolutionary coldspots (circles with unidirectional or no selection). After Thompson (2005).

coadaptation requires a particular level of gene flow. If there is too much mixing then adaptive alleles will be swamped by the flow of new nonadaptive alleles. On the other hand, if there is too little mixing then particular alleles will go to fixation and the coevolutionary process will grind to a halt, pending new mutations (Nash 2008).

The GMTC thus accounts for situations in which the intensity and direction of reciprocal selection and coadaptation vary across geographic regions (Hanifin et al. 2008). For example, Japanese camillia weevils (*Curculio camelliae*), which are seed predators, have remarkably long mouthparts (rostra) that are used to chew holes in camillia fruits (*Camillia japonica*) (plate 1c). If a female weevil can chew a deep enough hole, she deposits eggs into seeds at the center of the fruit with her ovipositor. To avoid damage to their seeds by weevils, the plants are under selection to increase the thickness of their fruits, assuming they have sufficient resources. The weevils are under reciprocal selection to increase the length of their mouthparts. However, reciprocal selection has not led to matching plant and weevil traits across all locations. Differences in ambient temperature complicate the pattern because the plants are capable of making thicker fruits at warmer latitudes, compared to colder latitudes. Phylogeographic history also plays a role, with historically subdivided populations responding differently to reciprocal selection (Toju 2009, 2011; Toju and Sota 2006).

The GMTC is consistent with the fact that, at a given location, reciprocal selection and coadaptation can also vary among years (Jokela et al. 2009). Importantly, absence of reciprocal selection between species at a given time or place does not necessarily mean that those species are *not* coadapting. It is increasingly clear that, for most systems, understanding coevolutionary dynamics requires data from more than one geographic location, and from more than one year.

A central feature of the GMTC is dispersal biology. Dispersal can be active, as in the case of juveniles emigrating from a natal breeding site, or passive, as in the case of wind-borne pollen (Clobert et al. 2001, 2012; Perrin 2009). Regardless of the mechanisms involved, dispersal is usually inversely correlated with geographic distance. Dispersal limitations constrain gene flow and contribute to random genetic drift. Dispersal also influences selection by generating new combinations of species or genotypes. The joint action of dispersal and selection provide much of the ecological context for micro- and macroevolutionary change (Reznick and Ricklefs 2009; Sobel et al. 2010).

Experimental coevolution

In some systems, it is possible to manipulate and observe coadaptation over many generations. This approach, called "experimental coevolution," uses lab-cultured microbial or other model systems (Brockhurst and Koskella 2013). Two or more microbial species are cultured together, while comparing 1) coadaptive responses between the species, or 2) the responses of lineages allowed to coevolve, compared to control lineages not allowed to coevolve. Some systems can be cryopreserved, allowing "time shift" experiments in which hosts are infected with parasite lineages from different points in time. Parasites can also be exposed to hosts from different points in time.

An example of this approach is to measure how parasite virulence evolves against a constant host background (Ebert 1998; Gaba and Ebert 2009). Alternatively, hosts from the "past" or "future" are experimentally infected with identical parasites to determine how host defense evolves against a constant parasite background. Other experimental coevolution studies use hosts and parasites from different geographic regions to test for local adaptation; the fitness of hosts infected with sympatric versus allopatric parasites is compared. Experiments of this kind show that local adaptation varies considerably among systems and conditions (Brockhurst and Koskella 2013). Local adaptation can be correlated with environmental productivity or environmental heterogeneity (Lopez-Pascua et al. 2012). Local adaptation can also be correlated with life history features and rates of parasite dispersal (Greischar and Koskella 2007; Hoeksema and Forde 2008; Vogwill et al. 2010).

Experimental coevolution studies have contributed greatly to our understanding of coadaptive responses to selection, the genetics of coadaptation, and limits to coevolutionary change (Buckling et al. 2009). The experimental coevolution approach also holds promise for work on the influence of coadaptation on diversification. For example, Bérénos et al. (2012) recently showed that experimental coadaptation between flour beetles (*Tribolium castaneum*) and their microsporidian parasites leads to increasing reproductive isolation among host lines within 17 generations.

Coadaptive diversification

Coadaptation can trigger ecological speciation (Nosil 2012), which is the evolution of reproductive isolation between populations exposed to different selective regimes. Indeed, the geographic mosaic theory of coevolution predicts this outcome (Thompson 2005). Coadaptive diversification is called "diversifying coevolution" by some authors, and "coevolutionary diversification" by others (box 1.1). Ehrlich and Raven (1964) proposed that the process leads to the reciprocal diversification of interacting groups. This process of "coadaptive codiversification" (fig. 1.6) has been a subject of intense interest and research for decades (Mitter et al. 1988; Jaenike 1990; Farrell 1998; Farrell and Sequeira 2004; Singer and Stireman 2005; Futuyma and Agrawal 2009; Coley and Kursar 2014). In chapter 12 we provide a framework for how selection and reciprocal selection can influence the diversification and codiversification of interacting groups.

Althoff et al. (2014) suggested three necessary criteria for a rigorous demonstration of diversifying coevolution: 1) coevolution (coadaptation) facilitates divergence between the populations of at least one of the coevolving species; 2) coevolving traits influence reproductive isolation, either directly or indirectly, among diverging populations; 3) coevolving lineages have higher net rates of diversification than non-coevolving lineages. Testing all three of these criteria requires a combination of work in ecological time (preferably experimental in nature) and macroevolutionary time (comparison of phylogenetic patterns of diversification).

One of the best examples in which these approaches have been combined is the elegant work of Benkman and colleagues on red crossbills (*Loxia curvirostra*) and the Rocky Mountain lodgepole pines (*Pinus contorta*), on which they feed (Smith and Benkman 2007). In areas of the western United States where crossbills are the dominant seed predators, they select for larger, thicker-scaled cones that protect pine seeds. Increased cone size exerts reciprocal selection on crossbills for increased beak size. However, in

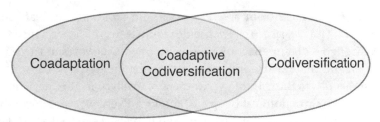

FIGURE 1.6. Interface of coadaptation and codiversification. Coadaptation is the joint microevolution of interacting species in response to reciprocal selection. Codiversification is the correlated diversification of interacting lineages. Coadaptive codiversification is the correlated diversification of interacting lineages in response to reciprocal selection (box 1.1).

areas where red squirrels (*Tamiasciurus hudsonicus*) are the dominant seed predator and outcompete crossbills, the cones are smaller and the birds have smaller beaks. Because populations of crossbills with different beak sizes do not interbreed readily, crossbills in regions with and without squirrels have undergone speciation. Thus, reciprocal selection and coadaptation between crossbills and pines, mediated by squirrel seed predators, facilitates reproductive isolation and diversification of crossbills (Benkman et al. 2010). Note that in this case coadaptation causes the diversification of crossbills, but not the pines, which are ancient lineages (He et al. 2012). This is an excellent example of coadaptive diversification (box 1.1).

Back in time

Questions about diversification quickly scale to longer spans of macroevolutionary time. In principle, it is straightforward to document reciprocal selection and coadaptation in ecological time (which is not to say it is easy). By comparison, demonstrating the influence of coadaptation over macroevolutionary time is daunting. Merely intuiting macroevolutionary change is difficult. By analogy, when we gaze upon stars in the next galaxy, we see light that has traveled for millions of years, yet we experience that light only in current time. In much the same way, we experience species and their interactions in ecological time, even though we know they are descendants of ancestors and processes that are millions of years old. Given the vast time periods involved, is it really possible to learn much about macroevolution from work restricted to ecological time?

One source of information is the fossil record, which provides snapshots of ancient interactions. For example, fossil pollen associated with insect fossils shows that insects were engaged in mutualistic interactions with angiosperms more than 100 million years ago (Hu et al. 2008). Ancient parasitic interactions are also apparent in the fossil record; fossil angiosperm leaves reveal damage from 97-million-year-old lepidopteran leaf miners (Labandeira et al. 1994, 2014). Even older fossils—from more than 356 million years ago—document interactions between liverworts in the Middle Devonian and their arthropod herbivores (Labandeira 2013). These fossils suggest that oils in the cells of liverworts were involved in chemical defense, similar to the terpenoid oil bodies of modern liverworts. Herbivory may have been widespread in the Devonian, and it may have played a significant role in structuring early terrestrial ecosystems.

Fossils also reveal the antiquity of parasites in animal hosts. Among the oldest evidence is the discovery of tapeworm eggs in 270-million-year-old shark feces (coprolites) (Dentzien-Dias et al. 2013). Even dinosaurs had parasites. Wolff et al. (2009) suggested that the severe erosive lesions on the mandibles of the *Tyrannosaurus rex* fossil "Sue" are consistent with trichomonosis, a disease caused by *Trichomonas gallinae*–like protozoan parasites (fig. 1.7). Indeed, the authors suggest that this parasite may have caused Sue's demise.

Amber is another useful source of information about the antiquity of interspecific interactions (DeVries and Poinar 1997). Schmidt et al. (2012) described gall mites trapped in 230-million-year-old Triassic amber. Poinar (2009) described a 100-million-year-old termite with a ruptured abdomen containing ten new species of flagellates, as well as a new genus of amoeba. Because termites rely on intestinal protozoans to help digest cellulose, this discovery is among the oldest evidence of mutualistic interactions between animals and microbes. Sources of information about ancient diets are not limited to organisms trapped in amber. Wappler et al. (2004) described a 44-million-year-old fossil bird louse containing its last meal of feather barbules. The fossil louse is remarkably similar to modern-day waterfowl lice (plate 3).

Examples like these provide information about the minimum ages of interactions; however, they cannot tell us much about coadaptive dynamics, per se. In some cases, it is possible to work out different stages of coadaptive change. Fossil time series for the marine bivalve *Merenaria* and its predatory whelk *Sinistrofulgar* show correlated increases in shell thickness, consistent with coadaptation (fig. 1.8a; Dietl 2003). Whelk predation selects for an increase in bivalve thickness, which may reciprocally select for an increase in whelk size, leading to coadaptation of these traits. An alternative interpretation is that increases in whelk size and bivalve thickness are joint responses to "top-down" selection imposed by durophagous crabs that prey upon whelks (fig. 1.8b; Dietl and Kelly 2002). If so, then the increases in shell thickness of the whelk and bivalve are due to unidirectional selection imposed by crabs, rather than an arms race between whelks and bivalves.

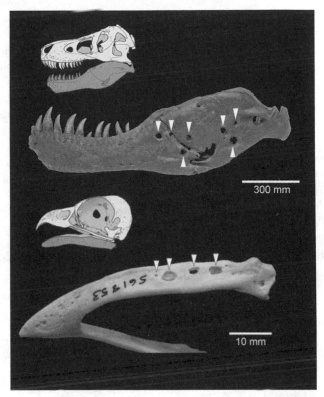

FIGURE 1.7. Sickly Sue. Trichomonosis-type lesions (arrows) in the left mandibular ramus of the *Tyrannosaurus rex* fossil known as "Sue" (upper illustrations), compared to lesions in the mandible of a modern osprey (*Pandion haliaetus*) with trichomonosis (lower illustrations). The distribution, size, and shape of the *T. rex* lesions are similar to those of the osprey, suggesting that Sue had *Trichomonas*-like protozoan parasites. After Wolff et al. (2009).

This second scenario is an example of unidirectional escalation, rather than coadaptation.

There are few examples of coadaptive dynamics in the fossil record, for several reasons. First, the fossil record is notoriously patchy, and the fossil record of interactions is even patchier. Second, the fossil record is limited to morphology, making it difficult to infer performance, unless inferences can be made from extant relatives (Dietl 2003; Dietl et al. 2010). Third, it is difficult to know the exact selective agents responsible for particular changes in fossil series. Just as in the case of extant communities, past assemblages were species rich, making it difficult to known exactly who interacted with whom, and for how long (Vermeij 1983, 2008). These difficulties notwithstanding, the fossil record of interspecific interactions continues to

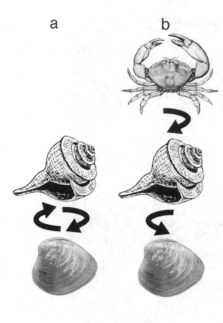

a   b

FIGURE 1.8. Direction of selection pressures in coevolution vs. unidirectional escalation of predator-prey interactions. (a) depicts a coevolutionary arms race in response to reciprocal selection between the predatory whelk *Sinistrofulgar* and its bivalve prey *Merenaria*. *Sinistrofulgar* uses its shell lip like a can opener to wedge and chip open the *Merenaria*, which is thought to select for thicker shells in *Merenaria*. Reciprocally, thickening of *Merenaria*'s shell selects for thickening of *Sinistrofulgar*'s shell, which can break while opening the thickest bivalves. *Merenaria* thus represents a "dangerous" prey item that can touch off a predator-prey arms race (Brodie and Brodie 1999). (b) "Top down" selection by a durophagus crab on the *Sinistrofulgar* whelk, leading to an increase in whelk size, which in turn selects for an increase in bivalve thickness. This cascade of selective effects is an example of unidirectional escalation, rather than coadaptation. After Dietl and Kelly (2002).

improve dramatically as additional specimens are discovered and studied (Vermeij 2008; Labandeira 2013).

Codiversification

Interactions often persist over macroevolutionary time, leading to patterns of codiversification (box 1.1.) at the *phylogenetic interface* between the interacting groups (fig. 1.9). Cophylogenetic methods (de Vienne et al. 2013) are used to document the intimacy and duration of these interactions over macroevolutionary time. When coupled with natural history information, cophylogenetic approaches provide a robust framework for exploring the different processes influencing codiversification.

Phylogenies of interacting groups can be more or less congruent (fig. 1.10a,b). The process generating congruence is *cospeciation*, which is the concomitant speciation of ecologically interacting groups (box 1.1). If every speciation event in one group is accompanied by a speciation event in the other group, and if no species change their associations or go extinct,

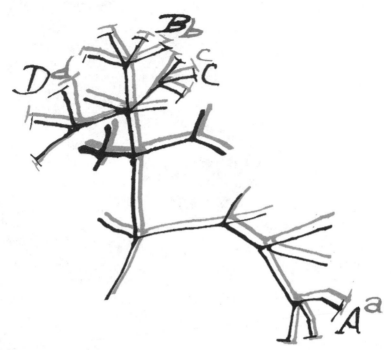

FIGURE 1.9. Phylogenetic interface of coevolution (with apologies to Darwin). In this example, Darwin's tree is completely congruent with another tree imagined by the authors. Both the topology and branch lengths are identical, making the phylogenies mirror images. After Darwin's (1837) Notebook B on Transmutation of Species.

then the phylogenetic interface will be congruent (fig. 1.10a). More often, however, phylogenies show partial congruence, or no congruence. Congruence is a pattern, whereas cospeciation is a process that generates this pattern. The term cospeciation is sometimes used in reference to patterns of congruence, but we find this confusing.

It is important to realize that cophylogenetic congruence can also arise for reasons having nothing to do with cospeciation. In the case of shared vicariance, for example, rivers or other barriers can split the range distributions of different groups along the same lines, leading to cophylogenetic congruence (C. I. Smith et al. 2008; Althoff et al. 2014). Note that codiversification in this case is not the result of ecological interactions between the groups. It happens simply because the groups share the same geographic distributions. Similarly, if two groups colonize a succession of islands in the same sequence, their phylogenies may be congruent in the absence of cospeciation. As yet another example, if parasites colonize a succession of hosts in the same sequence the hosts originally speciated, then the host and

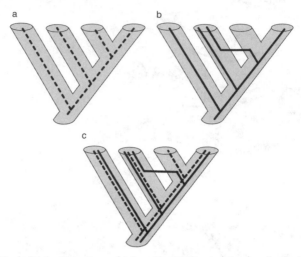

FIGURE 1.10. Cophylogenetic patterns for hypothetical hosts (pipes) and parasites (lines): (a) congruent phylogenies; (b) incongruent phylogenies because of a host switch; (c) ecological replicate approach, in which ecologically similar (phylogenetically independent) parasites are compared to their shared host group.

parasite phylogenies may be congruent, but without cospeciation. Cophylogenetic congruence without cospeciation has been called "pseudocospeciation" (Hafner and Nadler 1988).

Phylogenies that show extensive congruence are uncommon (de Vienne et al. 2013). It is more common for groups to show some evidence of cospeciation, mixed with other macroevolutionary events that "erode" congruence (fig. 1.10b). These events include host switching, lineage duplication, extinction, cohesion, and sorting events, all of which we review in chapter 9. Generally speaking, congruence is related to the ecological intimacy of interacting groups. One end of the spectrum is represented by mitochondria, chloroplasts, and other eukaryotic organelles that have evolved from free-living prokaryotic ancestors and are now part of the "host." The other end of the spectrum consists of much less intimate interactions, like those between generalist herbivores and their host plants. Most interactions lie between these extremes.

Congruent phylogenies can help confirm the duration of associations over macroevolutionary time. In a sense, they are analogous to the fossil record in allowing one to infer the longevity of interactions. Dating interactions requires that ancestral nodes of phylogenies be calibrated against known-age fossils to "set" a molecular clock. Although congruent phylogenies can provide evidence for the duration of associations, they do not providc cvidence for reciprocal selection or coadaptation. Additional data, like those from ecological experiments, are needed to put coadaptive flesh on the cophylogenetic bones. As discussed above, coadaptive diversification occurs when coadaptation between interacting lineages leads to diversification. Testing for coadaptive diversification requires an approach that integrates ecology and history (Weber and Agrawal 2012), a major theme of this book. Host and external parasite systems are excellent candidates for integrative work of this type.

Permanent parasites and ecological replicates

Many, if not most, coevolved interactions involve parasites. According to Price (1980), "A parasite is an organism living in or on another living organism, obtaining from it part or all of its organic nutriment, commonly exhibiting some degree of adaptive structural modification, and causing some degree of real damage to its host."

According to this definition, parasitism is the most common lifestyle on earth (Price 1980; Windsor 1998; Lafferty 2010). Parasites are remarkably diverse; they include microbial pathogens of animals, plant rusts, endoparasitic worms, herbivorous insects (plate 1a-c), vertebrate ectoparasites (plate 1d-f), and many others (Poulin and Morand 2004; Agrios 2005). By even the most conservative estimates, at least one third of all eukaryotic animals and plants are parasites (fig. 1.11).

Unfortunately, most organisms—including parasites—are intractable for work at the interface of micro- and macroevolution. Darwin's (1859) tangled bank may be interesting to contemplate, but it can be a tough place to run a controlled experiment (but see Morris et al. 2004). Although complex life cycles and communities are fascinating, they arc difficult to study, especially under natural conditions. Lab-based models, like those used in experimental coevolution studies, solve some complications by controlling for unknown selective agents and sources of environmental variation; however, they do so at the price of loss of real-world environments (Brockhurst and Koskella 2013).

Permanent parasites, such as lice (plate 2), provide an intermediate. Such parasites pass all stages of their life cycle on the body of the host (Rothschild and Clay 1957; Clayton and Moore 1997; Lehane 2005). This fact makes it possible to study them under natural conditions on captive hosts under controlled lab settings. Permanent parasites thus represent a melding of the laboratory and field. Some groups have phylogenies that are congruent with the host's phylogeny, allowing one to reconstruct interactions over macroevolutionary time. This combination of micro- and macroevolutionary tractability makes permanent parasites outstanding candidates for

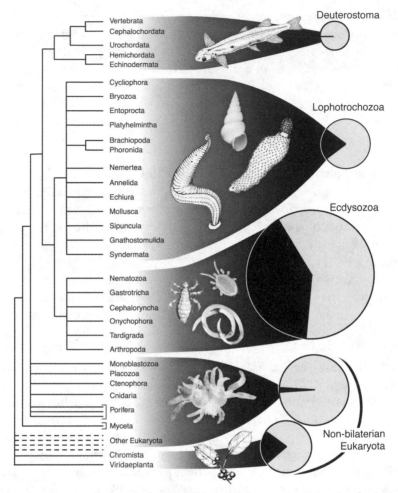

FIGURE 1.11. Diversity of parasitic taxa (black pie regions) versus free-living taxa (gray pie regions) among the different major lineages of eukaryotic organisms. Of the ca. 1.8 million described species, at least one-third are parasitic. The illustrated taxa are examples of parasites from each of the major clades (top to bottom, left to right: catfish, leech, snail, thorny-headed worm, sucking louse, mesostigmatid mite, nematode round-worm, jellyfish, mistletoe.) Pie diagrams show the rough proportion of taxa that are parasitic within each major group. The figure (by S.E.B.) is based on data from de Meeûs and Renaud (2002).

work at the interface of ecology and history. In essence, they buffer the agony and the ecstasy of evolutionary complexity.

Some hosts have multiple lineages of permanent parasites, making it possible to work on community interactions under natural conditions in the lab. In certain cases, these communities include "ecological replicates,"

FIGURE 1.12. Example of ecological replicate lice on mourning doves (*Zenaida macroura*). Oval-shaped body lice (*Physconelloides zenaidurae*) live on the bird's abdomen, where they feed on the downy regions of feathers. Cigar-shaped wing lice (*Columbicola macrourae*) spend most of their time on the flight feathers of the wings or tail, which are too coarse for the lice to eat. Like body lice, wing lice feed on the downy regions of abdominal feathers. Wing and body lice both attach their eggs to feathers and complete all stages of the life cycle on the body of the host. Photograph by Raees Uzhunnan.

which are ecologically similar parasite lineages that are phylogenetically independent (fig. 1.10c; Johnson and Clayton 2003b). Ecological replicates are a template for exploring factors that influence codiversification because their macroevolutionary histories with the shared host group can be compared. If the same cophylogenetic approach is used for each of the parasite groups, then *differences* in their cophylogenetic histories are likely to be robust, regardless of biases imposed by different types of cophylogenetic analysis or other factors. One can then design ecological experiments to test possible factors responsible for different macroevolutionary patterns. This ecological replicate approach works best for parasites that have similar life histories, but differ in features that may shed light on patterns of codiversification (e.g., Harbison and Clayton 2011).

In chapters 5–8 and 10 we include work that we have carried out with a pair of ecological replicate feather lice that live on columbiform birds (pigeons and doves). The pair consists of body lice and wing lice (fig. 1.12), which are members of the same family (Philopteridae). They have the same life cycle and both feed on the downy regions of the host's abdominal feathers. Despite their ecological similarity, the two groups show differences in both micro- and macroevolutionary time. Body lice are more host specific than wing lice (Johnson et al. 2002b), and their phylogeny is more similar to that of the host than is the phylogeny of wing lice (Clayton and Johnson 2003).

Comparison of these two groups allows one to address a variety of questions. Do they differ in host specificity because they differ in dispersal ability, or because they differ in establishment ability once on a novel host, or both? What are the consequences of the different mechanisms used by the lice to escape from host defense? Do body and wing lice compete and, if so, what are the consequences? Do body and wing lice interact with other members of the parasite community in similar ways, or differently, and how does this affect coevolution? These and other questions can be addressed, as we show in the coming chapters.

To provide background for answering these questions and others related to lice, we begin with an overview of the biology of lice (chapter 2), followed by a review of the effects of lice on their hosts (chapter 3).

2 BIOLOGY OF LICE: OVERVIEW

> I think pacifists might find it helpful to illustrate their pamphlets
> with enlarged photographs of lice. Glory of war indeed!
> —Orwell 1938

The purpose of this chapter is to provide a brief introduction to the biology of lice, particularly for readers with little knowledge of their biology. Some topics are covered in further detail in later chapters.

Lice are wingless insects that are obligate ectoparasites of birds and mammals. They have small eyes (or no eyes) and sense organs in their mouths and on their antennae (Clay 1970; Crespo and Vickers 2012). Lice are permanent parasites that complete all stages of their life cycle on the body of the host (fig. 2.1). They attach their eggs to fur or feathers with glandular cement. Both the eggs and hatched lice require warm, humid conditions near the host's skin to live (Marshall 1981a). Lice are sensitive to temperature and use thermal gradients to orient between different regions of the host's body (Harbison and Boughton 2014). Most lice are so dependent on the warm, humid conditions near the host's skin that they cannot survive for more than a few hours or days off the host (Tompkins and Clayton 1999).

Lice were traditionally divided into the two orders Anoplura, or "sucking lice," and Mallophaga, or "chewing lice." Because sucking lice were evolutionarily derived from within chewing lice (Lyal 1985; Johnson et al. 2004), modern classifications combine the two groups of lice into the single order Phthiraptera (table 2.1; fig. 2.2). Sucking lice are parasites of placental mammals; they have piercing-sucking mouthparts (not shown) that emerge from the anterior tip of a narrow head (plate 2a). Chewing lice, which are more diverse, are parasites of birds, as well as both placental and marsupial mammals (Price et al. 2003; Durden and Lloyd 2009). Chewing lice have broad heads with mandibulate mouthparts that are used to bite or scrape the host's integument (plate 2b, c). Chewing lice were traditionally called biting lice (Clay 1949a), but this is inaccurate because sucking lice also bite their hosts. Chewing lice consume feathers, skin debris, and secretions; some species also consume any blood on or near the skin's surface (Marshall 1981a; Lehane 2005; Mey 2013). Other chewing lice feed primarily on the downy regions of abdominal feathers (chapter 3). A few species of avian

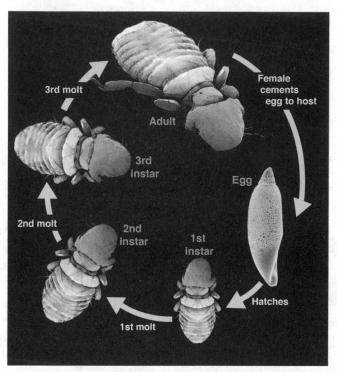

FIGURE 2.1. Lice have a hemimetabolous life cycle that begins with a large egg glued to the hair or feathers of the host. Three nymphal instars are completed after the egg hatches, followed by the adult stage. Eggs require 4–10 days to hatch, and each nymphal stage requires 3–12 days. Adult lice live 1–2 months on the host, with females producing 1–10 eggs per day (Marshall 1981a). The stylized life cycle shown here was created using a single scanning electron micrograph (SEM) of an adult *Myrsidea nesomimi* from a Galápagos mockingbird (*Mimus parvulus*) (SEM by V. S. Smith), and an egg of *Myrsidea invadens* from a common myna (*Acridotheres tristis*) (SEM by Gupta et al. 2009). Stages are roughly to scale, except the egg.

chewing lice have their mandibles modified into crude sucking mouthparts for feeding on subcutaneous blood (Clay 1949a; Nelson 1987).

Sucking lice (parasites of mammals)

Johnson and Clayton (2003a) reported a total of 532 species of sucking lice (table 2.1). About 20% of all mammal species are known to have sucking lice, with 70% of sucking lice associated with rodents (Kim 1985a,b; Light et al. 2010). Sucking lice are known to parasitize most orders of placental mammals, with the exception of cetaceans (whales and porpoises), sirenians (dugongs and manatees), xenarthrans (anteaters, armadillos, and

Table 2.1. Higher level classification of lice (Insecta: Phthiraptera)

Suborders & Families	Genera	Species	Hosts
[S]Anoplura (16 families)	49	532	mammals
[C]Amblycera			
Menoponidae	68	1,039	birds
Boopiidae	8	55	mammals[1]
Laemobothriidae	1	20	birds
Ricinidae	3	109	birds
Gyropidae	9	93	mammals
Trimenoponidae	6	18	mammals
[C]Ischnocera			
Philopteridae	138	2,698	birds[2]
Trichodectidae	19	362	mammals
[C]Rhynchophthirina			
Haematomyzidae	1	3	mammals

NOTE. Based on Johnson and Clayton (2003a); newer taxa not included.

[S] Sucking lice

[C] Chewing lice

[1] One genus (*Therodoxus*) occurs on birds (cassowaries).

[2] One genus (*Trichophilopterus*), which occurs on mammals (lemurs), is sometimes classified in its own family Trichophilopteridae.

---

sloths), pangolins, elephants, and bats (Durden and Musser 1994). Their absence from cetaceans and sirenians, which also lack chewing lice, is presumably because these marine mammals have sparse hair and, more importantly, do not come ashore. By way of comparison, seals host several genera of sucking lice that have remarkable adaptations for living in marine environments (see below).

The absence of sucking lice from armadillos and pangolins, both of which also lack chewing lice, may be related to their hard integuments; however, this cannot explain why sucking lice are also not found on anteaters or sloths (sloths do have chewing lice). The absence of sucking lice from elephants may be related to the fact that they host blood-feeding chewing lice in another suborder, the Rhynchophthirina, which is the sister group of sucking lice (fig. 2.2). Some mammals, such as pigs, do have both sucking and chewing lice.

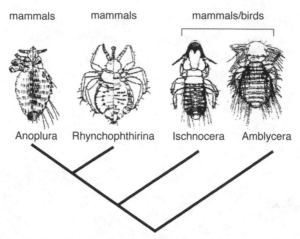

mammals   mammals   mammals/birds

Anoplura   Rhynchophthirina   Ischnocera   Amblycera

FIGURE 2.2. Subordinal level phylogeny for the insect order Phthiraptera (parasitic lice). Traditionally, Anoplura (sucking lice) had ordinal status, with the remaining three suborders (chewing lice) classified as the separate order Mallophaga.

Bats are completely devoid of lice, perhaps because of competition with an otherwise rich ectoparasite community that includes batflies, bugs, fleas, and mites (Marshall 1981a, 1982). Eichler (1936) and Wilson (1937) suggested that bats are free of lice because their body temperature drops steeply during hibernation; however, this hypothesis is inconsistent with the fact that tropical bats, which do not hibernate, are also free of lice. Furthermore, some other "deep" hibernating mammals, such as woodchucks (*Marmota monax*), do have lice (Hopkins 1949; Durden and Musser 1994). The patchwork distribution of sucking and chewing lice on the major lineages of mammals may reflect the relatively younger ages of mammal louse lineages, compared to bird louse lineages (Smith et al. 2011).

The blood diet of sucking lice is nutritionally incomplete. Recent studies confirm that sucking lice have endosymbiotic bacteria that can synthesize vitamin B and other supplements (Sasaki-Fukatsu et al. 2006; Perotti et al. 2007). For example, the human body louse (*Pediculus humanus humanus*) has a bacterium with a plasmid containing a unique arrangement of genes required for the synthesis of pantothenate, an essential vitamin deficient in the diet of the louse (Kirkness et al. 2010). Unfortunately, little is known about the physiological basis of symbiotic interactions between most sucking lice and their bacterial associates.

Chewing lice (parasites of mammals and birds)
The 4,397 total species of chewing lice reported by Johnson and Clayton (2003a) represent a paraphyletic group of three suborders: Amblycera,

Ischnocera, and Rhynchophthirina (table 2.1). Only about 12% of these species parasitize mammals. The rest parasitize birds, with every order and most families of birds known to have chewing lice (Price et al. 2003). Unlike sucking lice, which are restricted to placental mammals, chewing lice also parasitize marsupials, including kangaroos, quolls, wombats, and New World marsupials (Price et al. 2003). However, no lice have been recorded from monotremes (echidnas and platypi).

Like sucking lice, chewing lice have nutritionally incomplete diets. Feathers—a major component of the diet of many bird lice—are composed of keratins, which are hard to digest and have biased amino acid concentrations (Gillespie and Frenkel 1974). Not surprisingly, feathers are also limited in vitamins and co-factors (Waterhouse 1957). Like sucking lice, chewing lice have endosymbiotic bacteria (box 2.1). These bacteria provide lice with dietary supplements, such as B vitamins (Smith et al. 2010; Smith 2011). Chewing lice ingest some fungi and bacteria on the surface of the host's integument, but whether these contribute nutrients to the diet is unknown (Marshall 1981a). Fragments of mites in the alimentary tracts of both amblyceran and ischnoceran lice are common, but their nutritional value is also unknown (Pérez and Ateyo 1984). A few species of chewing lice feed on other lice, including members of their own species (Nelson and Murray 1971).

Amblycera, which tend to be rapid runners, will abandon dead or distressed hosts in search of a new host. In contrast, Ischnocera are so specialized for life on hair or feathers that they will not generally leave the body of the host, even if it is dead. Ischnoceran lice on birds are often called feather lice because they spend nearly all of their time on the feathers, and they consume mainly feathers (Johnson and Clayton 2003a). Some authors use "feather lice" in reference to bird lice, in general (Wappler et al. 2004; Hoeck and Keller 2012), but we find this confusing. We use "feather lice" only in reference to avian Ischnocera, the true feather specialists. Note that not all Ischnocera are feather lice; members of one family, the Trichodectidae, parasitize mammals (table 2.1).

The most widely known feather louse is probably *Strigiphilus garylarsoni*, described by Clayton (1990a) in honor of *Far Side* cartoonist Gary Larson (fig. 2.3). Larson featured the species in his book *Prehistory of the Far Side* (1989), with a printing of about 2.5 million copies, to date. Because Larson covered the end pages of each book with hundreds of miniature images of "his" louse, more than a billion copies of it have been published. In an interesting twist of fate, publication of Larson's book preceded publication of Clayton's paper by several weeks, meaning that Clayton (1990a) had to cite Larson (1989) as the first published illustration of *S. garylarsoni*!

BOX 2.1. The third tier: Endosymbiotic bacteria in lice

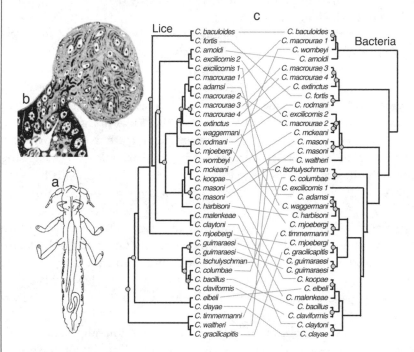

Endosymbiotic bacteria occur in a diversity of insects, including aphids, cockroaches, ants, whiteflies, weevils, and termites (Dale and Moran 2006). Endosymbionts such as *Buchnera*, from aphids, provide nutrients to their insect hosts, which have nutritionally incomplete diets (aphids feed on plant sap). The bacteria are transmitted vertically from the mother to her developing egg. *Buchnera* and aphids show repeated bouts of cospeciation, revealed by congruent phylogenies (Jousselin et al. 2009). Indeed, endosymbiotic bacteria and their hosts are among the best examples of cospeciation (de Vienne et al. 2013).

Many groups of lice also harbor endosymbiotic bacteria (Boyd and Reed 2012). Among the best known are those from *Columbicola* feather lice, including the lineage that lives in the rock pigeon wing louse, *C. columbae*. Ries (1931) first described these bacteria on the basis of light micrographs taken over 80 years ago. Images from his work are shown here: (a) line drawing of a male *C. columbae* showing parallel rows of bacteriocytes (in black) along the sides of the abdomen between the hypodermis and fat body; (b) micrograph of a female *C. columbae* with vertical transmission stages of endosymbiotic bacteria (stained black) penetrating the egg follicle (upper right) from a uterine horn (ovarial ampullae, lower left). Fukatsu et al. (2007) sequenced the 16S gene of *C. columbae* bacteria and, using in situ hybridization methods, confirmed that the bacteria live in specialized bacteriocytes and undergo maternal transmission (plate 4). Because *C. columbae* feeds largely on keratin-rich feathers, which are a nutritionally incomplete diet, it is likely that the bacteria play a role in dietary supplementation. Preliminary data support this hypothesis (Smith et al. 2010; Smith 2011).

The close association between *Columbicola* and their endosymbiotic bacteria makes them good candidates for cospeciation. However, a recent cophylogenetic analysis shows no evidence for cospeciation between these groups (Smith et al. 2013): (c) cophylogenetic reconstruction of *Columbicola* wing lice (left) and their bacterial endosymbionts (right). Phylogenies are based on maximum likelihood analyses of mitochondrial and nuclear genes for the lice (Johnson et al 2007), and 16S gene sequences for the bacteria (Smith et al. 2013). Thin lines show insect-bacteria associations (free-living bacteria and endosymbionts of other insects have been pruned from the tree). The trees are largely incongruent: the number of cospeciation events inferred from reconciliation analysis (grey bullets) is not more than expected by chance (P > 0.05). After Smith et al. (2013).

The star-like phylogeny of the bacteria includes many short internal branches, and long, highly variable branches leading to terminal taxa. This pattern suggests periodic acquisition/replacement of endosymbiont lineages by free-living bacteria over time, followed by accelerated rates of evolution once bacteria have made the transition to endosymbiosis. A free-living progenitor candidate has been identified on the basis of sequence similarity and phylogenetic position (see Clayton et al. 2012).

Ancient history

Until recently, Phthiraptera was the only insect order with no clear fossil record (Johnson and Clayton 2003a). This changed with publication of a magnificent fossil specimen from the crater of the Eckfeld maar near Manderscheid, Germany (Wappler et al. 2004). The 44-million-year-old specimen, christened *Megamenopon rasnitsyni,* is exceptionally well preserved (plate 3). Phylogenetic analysis of its morphology, compared to 47 extant taxa, revealed clear affinities to modern lice of aquatic birds. The fossil contains preserved feather barbules, confirming that *M. rasnitsyni* was a bird parasite. These results, coupled with paleoecological data, suggest that *M. rasnitsyni* parasitized a member of the Anseriformes (swans, geese, and ducks) or Charadriiformes (shorebirds). This discovery is one of the best examples of a fossilized host-parasite relationship.

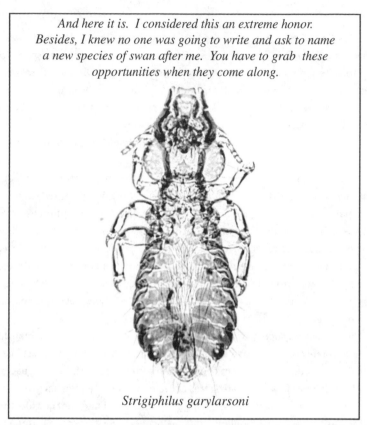

And here it is. I considered this an extreme honor.
Besides, I knew no one was going to write and ask to name
a new species of swan after me. You have to grab these
opportunities when they come along.

Strigiphilus garylarsoni

FIGURE 2.3. *Strigiphilus garylarsoni*, from the African owl *Ptilopsis* (formerly *Otus*) *leucotis*. This species was named in honor of cartoonist Gary Larson (Clayton 1990a), who featured it in his book *Prehistory of the Far Side* (1989; p. 171 reproduced here with permission of the publisher).

The antiquity of lice is further supported by the discovery of a 100+-million-year-old fossil of the booklouse family Liposcelididae (Grimaldi and Engel 2006). Members of this (nonparasitic) family are the closest relatives of parasitic lice (Lyal 1985; Johnson et al. 2004; Murrell and Barker 2005). The fossil booklouse, together with *M. rasnitsyni*, provides important calibration points for a recent phylogenetic analysis of 69 extant taxa of lice, which suggests that the major lineages (suborders) of parasitic lice began radiating long before the Cretaceous-Paleogene (K–Pg) boundary (fig. 2.4; Smith et al. 2011).

Interestingly, these results are inconsistent with the conventional hypothesis that most Cretaceous lineages of birds and mammals were

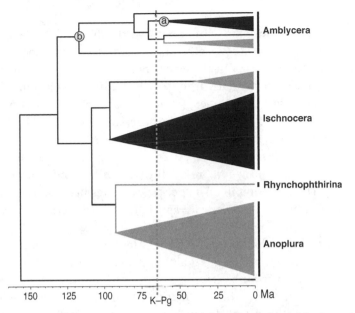

FIGURE 2.4. Simplified divergence time chronogram (in millions of years) for Psocoptera (booklice and barklice) and Phthiraptera (parasitic lice from birds, in black, and mammals, in grey) from Smith et al. (2011). Dates were established based on several calibration points including (a) *Megamenopon* (plate 3), a fossil louse that lived at least 44 million years ago (mya), (b) a fossil liposcelidid (booklouse) that lived at least 100 mya, as well as several other calibration points based on the host (bird and mammal) fossil record. Smith et al. (2011) showed that the lice began to diverge in the Cretaceous, long before an asteroid strike 65 mya caused mass extinctions. These data support the hypothesis that the bird and mammal hosts of these lice radiated in the Cretaceous, well before the K-Pg boundary. After Smith et al. (2011).

wiped out at the K-Pg boundary (Alvarez et al. 1980). This mass extinction hypothesis assumes that modern-day bird and mammal lineages are the result of a period of re-radiation from a handful of surviving lineages (Ericson et al. 2006; Wible et al. 2007). However, Smith et al. (2011) argued that the close association and specificity of parasitic lice with their hosts makes it unlikely that the mass extinction hypothesis is correct. Instead, the ancient radiation of lice supports the hypothesis that modern-day bird and mammal lineages began radiating long before the K-Pg boundary. Smith et al.'s paper is an interesting example of how parasites closely associated with host lineages can be used as "heirloom" parasites to reconstruct events in the evolutionary history of their hosts (Reed et al. 2004; Weiss 2009). For another example of this approach see box 9.2.

Host specificity

Many lice are host specific, with about two-thirds of the known species collected from just one species of host (Durden and Musser 1994a,b; Price et al. 2003). New species of lice and new host associations are discovered on a regular basis, particularly from poorly studied host groups, such as tropical birds (Clayton et al. 1996; Price and Johnson 2009; Kounek et al. 2011). Although most lice are host specific, some lice are known from several species, genera, or even families of hosts. An extreme example of a generalist is *Menacanthus eurysternus*, a songbird louse known from more than 175 bird species, representing 30 families and 2 orders (Price et al. 2003). This variation in host specificity makes lice useful models for exploring determinants of parasite specialization (Futuyma and Moreno 1988; Johnson et al. 2009).

The close association of lice with hosts in both micro- and macroevolutionary time also makes them good models for work at the interface of these coevolutionary time scales. Some groups of lice are closely associated with their hosts in macroevolutionary time, as revealed by phylogenies that are congruent with host phylogeny. Lice that show cophylogenetic congruence are host specific. However, just as permanence does not assure specificity, specificity does not assure congruence. Host specificity is an ecological measure that describes host use in microevolutionary time. By itself, specificity contains little or no macroevolutionary information (Clayton 1991a). Although some level of specificity is a precondition for cophylogenetic congruence, it is far from sufficient. For example, the bird louse genus *Brueelia* contains species that are host specific, yet their phylogeny shows no congruence with host phylogeny (Johnson et al. 2002a; Clayton et al. 2003a).

Understanding the relationship between specificity and congruence provides a window into the relationship between coevolution in micro- and macroevolutionary time. For example, are the factors that reinforce host specificity the same as the factors that lead to congruence? Are these factors influenced mainly by the dispersal limitations of lice, or does selection also play a role? If so, is the selection reciprocal and does it lead to host-parasite coadaptation? Does coadaptation reinforce codiversification?

Population biology

Rózsa (1997) showed that the abundance of lice among bird species is correlated with host body mass. He suggested that this relationship could be due to three factors: 1) large birds represent more resources for lice, 2) large birds represent more refugia from preening for lice, and 3) larger birds tend to live longer, providing more time for them to become infested with lice

(Rózsa 1997). Other studies have confirmed the relationship of host body size and louse abundance (Clayton and Walther 2001).

Within host species, lice tend to have aggregated distributions—that is, most individual hosts have few lice, while a few individuals have many lice (Eveleigh and Threlfall 1976; Fowler and Williams 1985; Rékási et al. 1997; Clayton et al. 1999). Rózsa et al. (1996) reported that aggregation is reduced on more social species of hosts, presumably because of increased horizontal transmission. Whiteman and Parker (2004) showed that aggregation is also reduced in more social populations of a single species of host, the Galápagos hawk (*Buteo galapagoensis*). Interestingly, in both studies the impact of host sociality on aggregation was greater for Ischnocera than Amblycera, perhaps because ischnoceran lice are less mobile than amblyceran lice. In short, host sociality appears to increase opportunities for horizontal dispersal marginally more in Ischnocera than Amblycera.

Mammals and birds have a variety of morphological, behavioral, and physiological defenses for combating lice. When these host defenses are impaired, lice populations increase dramatically. Stockdale and Raun (1960) showed that, on chickens with impaired preening ability, the chicken louse *Menacanthus stramineus* is capable of increasing from three to more than 12,300 lice in just six weeks! Host defense is not the only coevolutionary interaction that influences populations of lice. They are also influenced by competition with other lice (see chapter 6).

Most species of lice have even sex ratios (Marshall 1981b); however, some species have skewed sex ratios, usually with a female bias (Clayton et al. 1992). In an analysis of published data, Marshall (1981b) reported 31 of 50 species (62%) with significantly female-biased sex ratios, but none with male-biased sex ratios. In some cases males are rare, or absent altogether, suggesting parthenogenetic reproduction (Marshall 1981a,b; Westrom et al. 1976). Some populations of human head lice, with persistent female-biased sex ratios, carry the bacterial sex ratio distorter *Wolbachia pipientis* (Perotti et al. 2004). The causes and consequences of biased sex ratios in lice need further study.

Because the microclimate near the host's skin is relatively uniform, many lice are capable of breeding throughout the year (Askew 1971; Marshall 1981a). However, lice can still be influenced by non-host environmental factors, such as ambient humidity. Columbiformes (pigeons and doves) in more humid regions of the world have a higher prevalence of feather lice than Columbiformes in drier regions (fig. 2.5). The number of individual lice (intensity) is also higher in more humid regions (Moyer et al. 2002a; Malenke et al. 2011). Moreover, ambient humidity mediates the dynamics of interspecific competition in lice. Ambient temperature can also influence

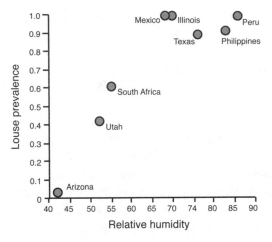

FIGURE 2.5. Mean prevalence of lice on Columbiformes (pigeons and doves) plotted against the mean annual relative humidity near the site of collection (n = 1,295 birds). Sampling localities were Tucson, Arizona, USA; Salt Lake City, Utah, USA; Free State and Mpumalanga, South Africa; Campeche, Mexico; Manteno, Illinois, USA; Weslaco, Texas, USA; Cagayan de Oro City, Philippines; and near Manu, Peru. After Moyer and Clayton (2004).

lice, even though they are closely attuned to the body temperature of the host. For example, intense solar radiation kills lice on sheep (Murray 1968), and on birds (Moyer and Wagenbach 1995).

Some lice have remarkable adaptations for withstanding the extreme habitats in which their hosts live. For example, sucking lice that infest Weddell seals (*Leptonychotes weddelli*) are capable of surviving long periods of immersion in the icy Antarctic waters. These lice have spiracles with a closing apparatus that conserves air and prevents flooding (Leonardi et al. 2012). Their abdomens are less heavily sclerotized than those of other lice, allowing cutaneous respiration from a bubble of air trapped by the spine-like setae (hairs) (Marshall 1981a). Weddell seal lice can survive 36 hours at −20°C (−4°F), and they show signs of heat stress at 25°C (77°F) (Murray et al. 1965). Since these lice require a temperature of 5–15°C to feed, they must wait until the host hauls out onto the ice, at which point they are capable of fully engorging themselves in less than five minutes (Marshall 1981a).

Successful transmission to new hosts is one of the greatest challenges faced by any parasite. Many internal and external parasites use several species of hosts to complete their life cycles, and they have elaborate adaptations for ensuring transmission between host species (Poulin 2007; Goater et al. 2013). Transmission of permanent parasites, such as lice, is much more straightforward (see chapter 7). It occurs mainly during periods of direct contact between hosts, such as that between parent hosts and their offspring, or between mated individuals (Marshall 1981a). However, some lice are capable of long-distance phoretic dispersal on parasitic flies (see chapter 7 and plate 5).

Community ecology

Communities of lice vary considerably in diversity among host groups. Species richness ranges from one species of louse per host, as in the case of elephants and ostriches, to more than a dozen species, as in the case of tinamous (Neotropical terrestrial birds). For example, more than 20 species of lice are known from the little tinamou (*Crypturellus soui*), with up to 9 species having been collected from a single individual (Ward 1957). In the case of tinamous, which are one of the oldest lineages of birds on earth (Hackett et al. 2008), it is possible that the rich community of lice reflects more time for speciation, colonization, and back colonization events. However, this does not explain why the ostrich, another old lineage, has only one species of louse. Ostriches have a relatively uniform covering of feathers, compared to tinamous, which have distinct feather tracts that presumably support the more diverse louse community.

Vas et al. (2012) demonstrated strong correlations between the species richness of host taxonomic families and the number of genera of lice found on species in those families, both for birds and mammals. This pattern, which is known as Eichler's rule (Eichler 1942; Stammer 1957), is perhaps not surprising, given that most genera of lice are relatively host specific.

Host geographic distributions and population size also influence the taxonomic richness of lice. Hughes and Page (2007) showed that the species richness of lice on seabirds is positively correlated with host population size and, to a lesser extent, host geographic range. Hosts colonizing new geographic regions, such as invasive species, tend to have fewer parasites, including lice (Paterson et al. 1999; Torchin et al. 2003). MacLeod et al. (2010) showed that this is not necessarily because colonizing host individuals have fewer lice, but because the lice on those hosts may fail to establish long-term viable populations on the introduced host populations. Successful establishment is more likely on introduced host species with larger body sizes or population sizes, presumably because of the higher probability of sustained transmission among the individuals of such hosts.

Clayton and Walther (2001) used phylogenetically independent contrasts to test possible determinants of louse species richness among more than 50 species of Neotropical birds. Their analyses revealed no features of host morphology or ecology that are significantly correlated with louse richness. However, richness itself was a significant positive predictor of the mean abundance of lice on different species. Mean abundance was also correlated with host body size, and the degree to which the upper mandible of the bill overlapped the lower mandible. This bill "overhang" turns out to be an important feature of host defense, as we discuss in chapter 4.

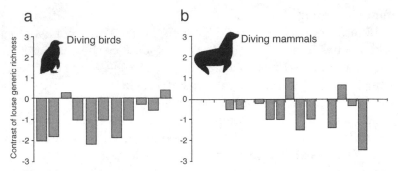

FIGURE 2.6. Phylogenetically independent contrasts of mean generic richness of lice on diving and non-diving sister clades of (a) birds, and (b) mammals. Negative values indicate that the diving species have fewer genera of lice than the non-diving species. After Felso and Rózsa (2006, 2007).

Host behavior also influences the taxonomic richness of lice. Felso and Rózsa (2006) showed that diving birds, such as osprey (Pandionidae), have fewer genera of lice than non-diving relatives, like hawks and eagles (Accipitridae) (fig. 2.6a). Felso and Rózsa (2007) showed a similar relationship for diving mammals, such as beavers (Castoridae), compared to non-diving relatives like squirrels (Sciuridae) (fig. 2.6b). The authors hypothesized that the negative effect of diving on louse richness is due to 1) the direct effect of water on lice, 2) something different about the structure of aquatic host feathers or fur, 3) something different about the grooming behavior of aquatic hosts, and/or 4) skin gland excretions of aquatic species that may influence lice.

Another intriguing correlate of the taxonomic richness of Amblycera (but not Ischnocera) is the cognitive ability of birds (Overington et al. 2009). Vas et al. (2011) reported significant positive correlations between the generic richness of avian Amblycera and the brain sizes and cognitive abilities of their hosts; these relationships were independent of host phylogeny and body size (fig. 2.7). The authors suggested several possible explanations: 1) more innovative birds use more diverse habitats, leading to more opportunities for host switching by lice; 2) more innovative birds are more social, and social birds have a higher abundance of lice, lowering the risk of louse extinction; 3) more innovative birds have more variable defensive behaviors, which could select for increased diversity in lice. Why the relationship between taxonomic richness and cognitive ability holds for Amblycera, but not Ischnocera, remains unknown.

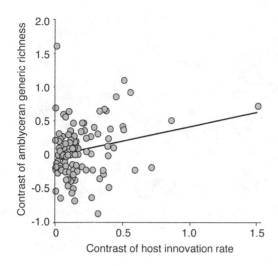

FIGURE 2.7. Phylogenetically independent contrasts of mean generic richness of amblyceran lice versus a measure of the innovation rate of different bird families (based, in part, on methods of searching, handling, or ingesting food). After Vas et al. (2011).

When it's personal: Human lice

The most thoroughly studied lice are those of humans (Weiss 2009). Although they may not admit it, some readers have been infested with head lice (*Pediculus humanus capitis*) (fig. 2.8a). Although they are even less likely to admit it, some readers have been infested with pubic lice (*Pthirus pubis*), also known as "crabs" (fig. 2.8b). The lead author of this book had a case of crab lice when he was a young graduate student. Since he was already studying lice, he effectively became his own field site, which is not recommended. He got the crab lice from his girlfriend, who claimed she got them from a towel.

Humans originally got crab lice, not from towels, but from the ancestors of gorillas a million or more years ago (see chapter 9). Crab lice and head lice are usually found in pubic hair and head hair, respectively. This is because they have tarsal claws adapted to coarse pubic hair or fine head hair (Reed et al. 2004). Crab lice are also known to inhabit other regions of coarse hair, such as the underarms (Longino, pers. comm.), as well as eyelashes, eyebrows, and chest hair (Rundle and Hughes 1993). Humans are the only members of the great apes with more than one genus of louse. The evolution of hairlessness, also unique to humans, facilitated the coexistence of two genera of lice by creating discrete microhabitats on a single host (box 9.2).

Neither head lice nor crab lice are known to vector pathogens or other parasites. In stark contrast, the human body or "clothes" louse (*P. h. humanus*) is a notorious vector of the bacteria responsible for three human

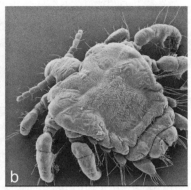

FIGURE 2.8. SEMs of (a) human head louse (*Pediculus humanus capitis*), and (b) human pubic louse (*Pthirus pubis*). Both species are sucking lice (Phthiraptera: Anoplura) that feed on human blood. Pubic lice, which are also known as crabs, have more robust claws than head lice. Crab louse claws fit around coarse pubic hair, but are too large to grasp finer head hair. The respective claw sizes of head and crab lice reinforce their microhabitat distributions. SEMs by V. S. Smith.

diseases: epidemic typhus (*Rickettsia prowazekii*), trench fever (*Bartonella quintana*), and relapsing fever (*Borrelia recurrentis*) (Busvine, 1976; Reed et al. 2011; Veracx and Raoult 2012; Allen 2014). These disease agents are among the greatest killers in modern human history. Of an estimated 30 million people who contracted epidemic typhus in World War I, about 3 million died (Lehane 2005). More than a million people in World War I also contracted trench fever (Brouqui 2011).

Body lice and the pathogens they vector may also have played a role in the retreat of Napoleon's army from Moscow in 1812. Raoult et al. (2006) analyzed DNA from soldiers exhumed from a mass grave in Vilnius, Lithuania. In addition to human DNA, the researchers recovered body lice DNA as well as DNA from two pathogens (*R. prowazekii* and *B. quintana*) vectored by the lice. The results of this study are molecular smoking guns suggesting that body lice may have prevented Napoleon from taking Moscow.

Although body lice and head lice do not appear to interbreed under natural conditions, fertile F_1 hybrids have been created under laboratory conditions (Mullen and Durden 2009). Recent population genetic, phylogenetic, and comparative genomic work also indicates that human head and body lice are not separate species, but ecotypes of a single species (Reed et al. 2004; Light et al. 2008; Olds et al. 2012). Nevertheless, debate concerning the species status of these lice, which has raged for decades, continues (Busvine 1978; Veracx and Raoult 2012).

A comparison of the transcriptomes of human head and body lice shows that they differ mainly in the expression levels of only 14 genes. Why, then, do head lice not vector pathogens? Previte et al. (2014) reported gene expression data consistent with the possibility that the immune systems of head lice are better at fighting early stages of infection with *B. quintana*; however, the specifics of the underlying mechanisms remain unknown. One thing is certain: should head lice start widely vectoring pathogens at some point, the public health implications could be frightening. Some data already suggest that head lice are capable of vectoring *B. quintana* and other pathogens under some circumstances, but the subject is controversial (Veracx and Raoult 2012; Bonilla et al. 2013).

As mere irritants, head lice are a severe problem for children and their parents throughout the world (Burgess 2004). It has been estimated that elementary school children in the United States miss 12–24 million days of school per year because of head lice (Roberts 2002). The problem is increasingly difficult to treat because head lice have evolved resistance to many of the most common pediculicides (Kwon et al. 2008). Fortunately, some nonchemical methods are effective for treating resistant head lice (Burgess 2004; Bush et al. 2011).

Rózsa and Apari (2012) proposed the intriguing hypothesis that head lice may actually have a positive effect on human health. Their hypothesis is based on the fact that head lice are common on children and relatively harmless, while body lice are more common on adults and represent a significant threat as vectors. The authors suggest that priming of the immune system by early infestation with head lice may help people resist body lice later in life. This, in turn, may have selected for host behavior that ensures transmission of head lice between mates, and from parents to their offspring. Interestingly, the authors point out that humans are the only primates to engage in regular head touching behavior between family members. Zinsser (1935) reviewed other possible benefits of lice to human health in his classic volume *Rats, Lice and History*.

3 EFFECTS OF LICE ON HOSTS

He has too many lice to feel an itch.
—Chinese proverb

A fundamental assumption underlying the process of co-adaptation is that interacting species have reciprocal effects on one another's fitness. Measuring effects of parasites on host fitness can be challenging because "third" variables, such as poor host nutrition, can drive spurious negative correlations between parasite abundance and host fitness. In this chapter we review data relevant to the effects of lice on host fitness, including aspects of host condition, survival, and reproductive success. We consider five nonexclusive categories: 1) effects of lice on domesticated hosts, 2) effects of lice on wild hosts, 3) effects of lice on mate choice, 4) costs of host resistance, and 5) host tolerance. Table 3.1 lists sources of additional information on the effects of lice on the fitness of both wild and domesticated mammals and birds.

Effects of lice on host fitness mediate selection for the evolution of host resistance, which we cover in chapter 4. Hosts may tolerate small infestations of lice, which we consider at the end of the current chapter.

Effects of lice on domesticated hosts

When present in large numbers, lice can have severe negative effects on domesticated hosts, occasionally resulting in their death. It is important to consider these effects because they are useful for understanding the potential effects of lice on wild hosts. However, data from domesticated hosts must be interpreted carefully, for several reasons. Because domesticated hosts are under relaxed natural selection, and because they tend to be housed under dense conditions, their lice may be unusually virulent. Furthermore, given increased opportunities for "straggling" of lice between host species living in close proximity, domesticated hosts may have lice and other parasites that are not found on their wild counterparts. For these reasons, and others, it is important to use extreme caution when interpreting interactions between domesticated hosts and lice.

Mammals

Sucking and chewing lice are major pests of ruminants, hogs, and other domesticated mammals (Campbell 1988; Durden and Lloyd 2009). In cattle,

Table 3.1. Reviews of the effects of lice on domesticated and wild hosts

Effect on host	Source
Direct effects of lice on domesticated mammals	Nelson et al. (1977), Marshall (1981a), Jones (1996), Price and Graham (1997), Lehane (2005), Durden and Lloyd (2009)
Direct and indirect effects of lice on wild mammals	Durden (2001, 2005)
Mammal lice as intermediate hosts and vectors	Durden (2001, 2005)
Direct effects of lice on domesticated birds	Nelson et al. (1977), Loomis (1978), Marshall (1981a), Arends (1997), Price and Graham (1997), Prelezov et al. (2006), Clayton et al. (2008)
Direct and indirect effects of lice on wild birds	Clayton et al. (2008)
Bird lice as intermediate hosts and vectors	Saxena et al. (1985), Clayton et al. (2008)
Influence of lice and other parasites on host mating and sexual selection	Clayton (1991b), Møller et al. (1999), Garamszegi (2005), David and Heeb (2009), Poulin and Forbes (2012)

large infestations of the sucking louse *Haematopinus eurysternus* cause severe anemia and death (Jones 1996). Domesticated water buffalo with large infestations of *H. tuberculatus* exhibit ruffled pelage, peeling skin, anemia, and pale mucous membranes, particularly in young animals (Da Silva et al. 2013a). The chewing louse *Bovicola bovis* is so annoying for cattle that patches of skin become raw as a result of excessive grooming; irritated cattle destroy fences, barns, and other structures by rubbing against them. Large numbers of *Bovicola equi* on horses cause intense itching, crusted lesions, hair loss, and scaly skin (Da Silva et al. 2013b). The chewing louse *Damalinia ovis* causes thickening of the skin in sheep (Britt et al. 1986), and the sucking louse *Linognathus pedalis* causes lameness in sheep (Lehane 2005). *L. vituli* can cause anemia and even death in young calves (Otter et al. 2003).

The sucking louse *Haematopinus suis* is extremely irritating to adult hogs, and it alters blood cell parameters of piglets (Davis and Williams 1986). Infested piglets show retarded development and slower growth than controls

(Kamyszek and Gibasiewicz 1986). *Haematopinus tuberculatus* infestations of sheep and goats are so irritating that they cause extensive biting and rubbing, which damages skin, wool, and hair. Goat lice can cause anemia and death, especially in juveniles (Durden and Lloyd 2009). Lice are also known to have negative effects on companion animals such as dogs and cats, as well as laboratory animals like mice and rats. On grooming-impaired lab mice, the louse *Polyplax serrata* causes fur loss, anemia, and even death (Clifford et al. 1967; Lehane 2005). Suffice it to say that, under domesticated conditions, mammal lice can have extremely negative effects on aspects of host fitness.

In addition to these direct effects of lice on mammals, both sucking and chewing lice may have indirect effects as vectors of viral, bacterial, and fungal pathogens, as well as endoparasitic worms (table 3.2). Lice also vector pathogens to the ultimate domesticated mammal, *Homo sapiens*, as discussed in chapter 2.

Birds

Chewing lice have severe effects on poultry and other domesticated birds. When present in large numbers, they cause extensive feather and skin damage, leading to dermatitis, itching, insomnia, and excessive preening and scratching. The amblyceran chicken louse *Menacanthus stramineus* causes significant irritation of the skin and, as a result, birds increase their preening rates (Brown 1974). *Menacanthus stramineus* also causes reductions in body mass and egg production (Nelson et al. 1977; Arends 1997; Prelezov et al. 2006).

The ischnoceran chicken louse *Cuclotogaster heterographus* causes severe restlessness and debility (Kim et al. 1973), and it can kill chicks outright (Loomis 1978). Poultry experimentally co-infested with four species of lice showed hematological changes (Prelezov et al. 2002) and histological changes in the skin, muscle, spleen, liver, lungs, kidneys, and small intestine, followed by the death of some birds (Prelezov et al. 2006). Lice have also been shown to have negative effects on turkeys and other domesticated fowl, as well as pet birds like parrots (Durden and Lloyd 2009).

Like mammal lice, bird lice serve as vectors and intermediate hosts of other parasites (table 3.1). The amblyceran louse *Trinoton anserium* transmits the swan heartworm *Sarconema eurycera*, which can kill swans (Seegar et al. 1976; Cohen et al. 1991). Ischnoceran feather lice also serve as intermediate hosts of filarid nematodes, which are ingested along with the lice when birds preen (Bartlett 1993). Viruses and bacteria have also been isolated from bird lice (table 3.2), but the extent to which lice actually play a role in microbial transmission is unknown.

Table 3.2. Parasites and pathogens isolated from lice

Vertebrate host	Louse	Parasite/Pathogen
Birds[1]		
African swift	[A]*Dennyus hirundinis*	[H]*Filaria cypseli*
American coot	[A]*Pseudomenopon pilosum*	[H]*Pelecitus fulicaeatrae*
Jungle fowl (domestic)	[A]*Eomenacanthus stramineus* (=*Menacanthus*)	[B]*Escherichia coli*
		[V]Eastern Equine Encephalomyelitis
		[B]*Pasteurella multicoida*
		[B]*Salmonella gallinarum*
		[B]*Streptococcus equinus*
	[A]*Menopon gallinae*	[B]*Escherichia coli*
		[B]*Ornithosis bedsoniae* (=*Chlamydophila*)
		[B]*Pasteurella multocida*
		[B]*Streptococcus equinus*
Marbled godwit	[A]*Actornithophilus limosae*	[H]*Eulimdana wongae*
	[I]*Carduiceps clayae*	[H]*Eulimdana wongae*
Mute swan	[A]*Trinoton anserinum*	[H]*Sarconema euryceru*
Whistling swan		
Red-necked grebe	[A]*Pseudomenopon* sp.	[H]*Pelecitus fulicaeatrae*
Whimbrel	[A]*Austromenopon phaeopodis*	[H]*Eulimdana bainae*
	[I]*Lunaceps numenii*	[H]*Eulimdana bainae*
Mammals[2]		
Water buffalo	[An]*Haematopinus tuberculatus*	[B]*Anaplasma marginale*
Cattle	[An]Cattle sucking lice	[B]*Anaplasma* spp.
		[B]*Rickettsia* spp.
	Cattle lice	[F]*Trichophyton verrucosum*
Dogs	[I]*Trichodectes canis*	[H]*Dipylidium caninum*
	[Bo]*Heterodoxus spiniger*	[B]*Anaplasma platys*
Goats	[An]*Linognathus stenopsis*	[B]*Anaplasma* spp.
		[B]*Rickettsia* spp.
Harbor seal	[An]*Echinophthirius horridus*	[H]*Dipetalonema spirocauda*
Hogs	[An]*Haematopinus suis*	[V]Pox virus
		[B]*Anaplasma* spp.
Humans	[An]*Pediculus h. humanus*	[B]*Rickettsia prowazekii*
		[B]*Borrelia recurrentis*
		[B]*Salmonella enteriditis*
		[B]*Bartonella quintana*

continued

Table 3.2. continued

Vertebrate host	Louse	Parasite/Pathogen
Mice (domestic)	[An]*Polyplax serrata*	[B]*Mycoplasma coccoides*
Rats (domestic)	[An]*Polyplax spinulosa*	[B]*Mycoplasma muris*
Rodents, rabbits	Rodent & rabbit lice	[B]*Francisella tularensis*
Voles	[An]*Hoplopleura acanthopus*	[B]*Brucella brucei*
Woodchuck	[An]*Enderleinellus marmotae*	[B]*Rickettsia typhi*

NOTE. In most cases the actual vector potential of the lice has not been tested.
[A]Amblycera
[An]Anoplura
[B]Bacteria
[Bo]Boopida
[F]Fungus
[H]Helminth
[I]Ischnocera
[V]Virus
[1]Clayton et al. (2008).
[2]Brown et al. (2005), Da Silva et al. (2013a), Hornok et al. (2010), Mullen and Durden (2009), Reeves et al. (2005).

When a good parasite goes bad: Effects of lice on wild hosts

At one time, the conventional wisdom was that lice on wild birds and mammals were "good" parasites with relatively little effect on host fitness (Rothschild and Clay 1957; Ash 1960; Marshall 1981a; Lehmann 1993). There were several reasons for this belief. First, lice are good candidates for the old idea that vertically transmitted parasites will evolve attenuated virulence (but see below). Second, feather lice are often viewed as harmless because they feed mainly on mature (dead) feathers and dead skin (Waage 1979; Hoi et al. 2012). Another factor contributing to the perception that lice are harmless is the fact that many individual wild hosts have few, if any, lice (Ash 1960; Marshall 1981a).

In recent years, evolutionary parasitologists have come to realize that the evolution of virulence is more complicated than originally thought; indeed, even vertically transmitted parasites can be quite virulent (Ewald 1994; Schmid-Hempel 2011). Moreover, virulence is not a static trait of a parasite or host, but changes in relation to environmental context. Parasites with little or no effect on host fitness in good years can be devastating in lean years. Moreover, small numbers of parasites on wild hosts are not indicative of the evolution of avirulence, but are more often a consequence of the evolution of efficient host defense. Host defense itself can be costly and reduce host fitness, as we consider later in this chapter. First we review data showing that lice can be quite detrimental to wild hosts.

Mammals

Lice can have both direct and indirect negative effects on wild species of mammals. For example, the sucking louse *Linognathus africanus*, which parasitizes sheep and goats, causes severe anemia and death when it infests wild mule deer (*Odocoileus hemionus*) (Brunetti and Cribbs 1971; Durden 2001). The sucking louse *Echinophthirius horridus* causes extensive fur loss in harbor seals (*Phoca vitulina*), and vectors a heartworm that can be fatal to seals (table 3.2; Skirnisson and Olafsson 1990). The chewing louse *Trichodectes canis* causes skin lesions and hair loss in gray wolves (*Canis lupus*), reducing their winter survival (Schwartz et al. 1983). *Trichodectes canis* is also an intermediate host for the tapeworm *Dipylidium caninum* (table 3.2). The elephant louse *Haematomyzus elephantis* causes severe dermatitis when present in large numbers (Raghavan et al. 1968).

The potential for stressful environmental conditions to exacerbate the effects of lice is illustrated by an account of two captive spider monkeys (*Ateles geoffroyi*) that died several weeks following importation by a roadside zoo in Missouri (Ronald and Wagner 1973). The monkeys roamed freely and interacted with hundreds of children. When necropsied, the diagnosed cause of death was anemia caused by large infestations of the blood-feeding louse *Pediculus mjobergi*, which is a parasite of New World monkeys (Durden and Musser 1994). This example illustrates how environmental conditions can trigger an increase in lice, with subsequent devastating effects. Presumably, the two unfortunate monkeys had little or no time to control their lice by grooming.

Birds

Many studies have tested for effects of lice on wild birds. Some studies have documented natural co-variation between lice and host condition, while others have tested effects of lice by experimentally manipulating parasite load (Clayton 1991a; Møller et al. 1999; Clayton et al. 2008). As an example of the first approach, Freed et al. (2008a,b) estimated the ectoparasite loads of several thousand Hawaiian birds over nearly two decades. In the final years of their study, the authors documented a dramatic increase in the prevalence of ectoparasites (mainly lice) among 11 species of forest birds (fig. 3.1). Although the reason for the increase remains unclear, it may have

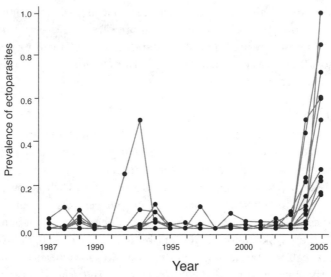

FIGURE 3.1. Estimated prevalence of ectoparasites (mainly lice) among 11 species of Hawaiian forest birds over 18 years. After Freed et al. (2008b).

been triggered by a decrease in host condition due to competition with the invasive Japanese white-eye *Zosterops japonicas* (Mountainspring and Scott 1985). Birds in poor condition are often less capable of mounting effective defenses against ectoparasites (see chapter 4), which may explain the increased ectoparasite loads of the Hawaiian birds. It is also possible that some of the lice switched to the native hosts from invasive white-eyes, but this remains unclear because the lice were not identified.

In other studies of natural co-variation between lice and host condition, lice are clearly a direct cause of poor host health. For example, in some years American white pelicans (*Pelecanus erythrorhyncus*) acquire heavy infestations of *Piagetiella peralis,* an amblyceran louse that lives in the pouch (Dik 2006). Juvenile pelicans with large infestations of these lice have badly inflamed mouth linings, are in poor condition, and can die. Although it is not known whether the lice are the main cause of host mortality, they are quite clearly bad for the birds (Samuel et al. 1982).

The gold standard for testing effects of lice and other parasites on host fitness is to experimentally manipulate parasite load and monitor downstream effects (McCallum and Dobson 1995). In one of the first such studies of wild birds involving lice, Brown et al. (1995) showed that cliff swallows (*Hirundo pyrrhonota*) with fleas, bugs, and two species of lice had lower survivorship than fumigated controls (Brown et al. 1995). Unfortunately, it was

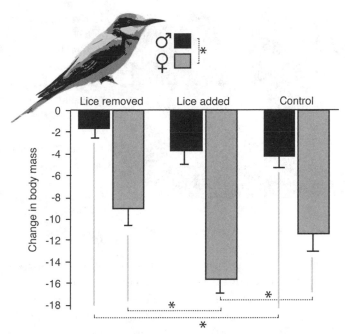

FIGURE 3.2. Effect of ischnoceran chewing lice on the body mass (grams) of breeding European bee-eaters (*Merops apiaster*). All birds lost mass over the course of the month-long experiment, which was conducted at the height of the breeding season. Female birds with experimentally increased louse loads lost more body mass than birds from which the lice were removed, or control birds from which the lice were temporarily removed, then immediately put back on the birds. Male birds from which lice were removed lost less body mass than male controls. Overall, female birds lost significantly more body mass than male birds. Asterisks (*) indicate significant differences (P < 0.05) between the treatments being compared (dotted lines). After Hoi et al. (2012).

not possible to tell how much of the survival effect was due to the lice, the fleas, the bugs, or some combination of the three.

More recently, Hoi et al. (2012) demonstrated a negative effect of two species of ischnoceran feather lice (*Meropoecus meropis* and *Brueelia apiastri*) on the body mass, hematocrit, and sedimentation rates of breeding European bee-eaters (*Merops apiaster*). After experimental manipulation of louse loads, male and female birds lost an average of 4% and 16% of body mass, respectively, over the course of the month-long field experiment (fig. 3.2). The fact that the lice affected female birds more than male birds may have been due to greater overall energetic demands on females than males during the breeding season.

Long-term experimental manipulations have also been used to test

FIGURE 3.3. Damage to rock pigeon (*Columba livia*) feathers from feeding lice (*Columbicola columbae* and *Campanulotes compar*). Abdominal contour feathers with (left to right) no damage, average damage, and severe damage. The downy region and barbules of the basal and medial regions have been consumed. From Clayton 1990b, by permission of *American Zoologist*.

effects of lice on wild birds. Clayton et al. (1999) conducted a one-year experiment with feral rock pigeons (*Columba livia*) infested with wing lice (*Columbicola columbae)* and body lice (*Campanulotes compar*) (Clayton et al. 1999). The two species of lice feed on the downy portions of the host's abdominal contour feathers (fig. 3.3). Both species were experimentally increased on birds using "bits" (fig. 4.4), which impair effective preening but not feeding (Clayton 1991a; Clayton and Tompkins 1995). Increases on bitted birds were well within the range of natural variation in louse load. Control birds also wore bits, but they were periodically fumigated to prevent their lice from increasing. Experimental birds showed significant reductions in feather mass (fig. 3.4a,b) and significant increases in thermal conductance and metabolic rate (Booth et al. 1993). The body mass (fig. 3.4c) and survival (fig. 3.4d) of experimental birds also decreased over time, compared to controls (Clayton et al. 1999). Thus, damage to downy feathers by feather-feeding lice caused energetic stress that significantly reduced body mass and survival of birds over the course of many months in the field.

Lice may also chew holes directly in the wing and tail feathers of birds, with important consequences for host fitness, but this hypothesis requires further testing (box 3.1).

Effects of lice on host mate choice
Hamilton and Zuk (1982) suggested that members of the choosier sex (usually females) exert selection for the evolution of secondary sexual traits that convey information concerning the parasite loads of prospective mates. Parasite-mediated mate choice of this kind has three adaptive benefits (Clayton 1991b, Møller et al. 1999). First, it can allow females to identify

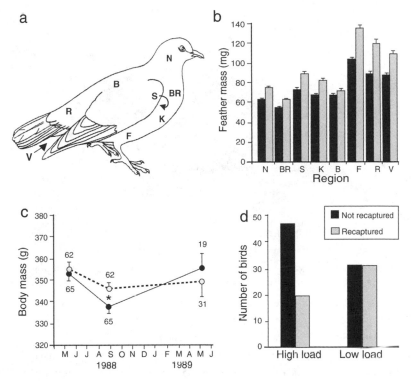

FIGURE 3.4. Effects of experimentally manipulated feather lice on feral rock pigeons (*Columba livia*) over the course of a year-long field experiment. (a) Regions of the body sampled for feathers; b) Mean (+se) mass of feathers from 58 high-load birds (black bars) and 56 low-load birds (gray bars). The value for each bird is the combined mass of the 10 longest feathers of a clump plucked from each of the body regions shown in a): *N* = nape, *BR* = breast, *S* = side (under wing), *K* = adjacent to keel, *B* = back, *F* = flank, *R* = rump, and *V* = ventral caudal tract.

There were significant effects of treatment, body region, and treatment × region (P < 0.001 in all cases), with high-load birds having less feather mass; c) Mean (±se) body mass of high-load birds (solid circles) versus low-load birds (open circles). Numbers above and below error bars are sample sizes. Only birds captured at the start of the study, and again 3 months later, were included. June 1989 data are for all birds recaptured 1 year after treatment. Significant difference (P < 0.05) is indicated by the asterisk; (d) Survival of high- versus low-load birds. The number of birds recaptured in June 1989 was compared to the number of birds treated the previous June (and recaptured and retreated in September 1988). The recapture rate of high-load birds was significantly less than that of low-load birds (P < 0.017). After Clayton et al. (1999).

BOX 3.1. Do lice make holes in flight feathers?

Kose et al. (1999) attributed holes in barn swallow feathers (*Hirundo rustica*), like the ones shown here (from Vas et al. 2008), to the amblyceran louse *Machaerilaemus* (formerly *Hirundoecus*) *malleus*. Male swallows with more holes return from the wintering grounds later than swallows with fewer holes, putting them at a breeding disadvantage (Møller et al. 2004a). Vas et al. (2008) questioned whether *M. malleus* causes such holes in the flight feathers of swallows. They suggested that a more likely culprit is the ischnoceran louse genus *Brueelia*. More recently, Vágási (2014) argued that it is not known whether any genus of louse really makes such holes in the flight feathers of birds. This debate has obvious implications for studies that use feather holes as proxies for louse load (Moreno-Rueda 2005; Moreno-Rueda and Hoi 2011). An experiment is needed in which birds are infested with lice and the condition of their flight feathers monitored over time, compared to uninfested controls. Until this experiment is carried out, the results of studies quantifying holes in feathers, rather than quantifying the lice themselves, must be evaluated with caution.

and choose males with "good" genes for parasite resistance, genes that are then likely to be inherited by the female's offspring. Second, healthy males can provide more assistance in rearing offspring, at least in species with bi-parental care. Third, choosing a parasite-free mate can minimize the risk of contagious transmission. This last benefit, variously labeled parasite avoidance (Borgia and Collis 1989), contagion indication (Able 1996), or transmission avoidance (Loehle 1997), is particularly relevant in the case of lice and other parasites that rely on host contact for transmission (but see Martinez-Padilla et al. 2012).

Sucking lice are known to influence mate choice in mammalian hosts. Female mice distinguish between the urinary odors of male mice parasitized by the sucking louse *Polyplax serrata*, versus mice that are not parasitized by these lice (Kavaliers et al. 2003b). Female mice avoid the odors of parasitized males in mate choice trials; however, females whose gene for oxytocin has been deleted (OT knockouts) no longer distinguish between males with and without lice (Kavaliers et al. 2003a, 2005). Thus, mate choice for parasite-free males can have a strong genetic component.

Lice also influence mate choice and sexually selected traits in birds. Feather damage from lice on feral rock pigeons causes reductions in male courtship display and mating success (Clayton 1990b). Thus, pigeon lice have a negative effect on both the reproductive and survival components of host fitness (see above). Spurrier et al. (1991) tested the effect of lice on the mating success of male sage grouse (*Centrocercus urophasianus*). They showed that amblyceran lice create visible haematomas on the air sacs of males, and that females avoid mating with males that have experimentally simulated haematomas. Lice may also influence the quality of sexually selected traits in barn swallows, such as tail length (Møller 1991; Saino and Møller 1994), or the duration of courtship song (Garamszegi et al. 2005). Unfortunately, most of the barn swallow studies have used counts of holes in flight feathers, which may be unreliable (box 3.1).

Another example of the influence of lice on mate choice involves the amblyceran louse *Myrsidea ptilonorhynchi*, which is a parasite of satin bowerbirds (*Ptilonorhynchus violaceus*) in Australia. Male birds construct bowers, which are elaborate structures made of twigs and decorated with colorful objects, such as flowers and berries. Females stand near the bowers and observe males engaged in courtship display. If a male is acceptable to the female, she copulates with him in the bower (Doucet and Montgomerie 2003a,b). Borgia et al. (2004) showed that, among males with bowers, females prefer individuals with the fewest lice. These lice, many of which are found on the host's head, have white eggs that contrast sharply with the dark blue plumage of the males. Females may conceivably assess parasite load by scrutinizing the facial feathers of males for louse eggs during courtship (Borgia and Collis 1989). In an independent study at a different location, Doucet and Montgomerie (2003a,b) showed that males with high-quality bowers had fewer lice, and that bower quality explained more than 50% of the variation in louse loads among male birds.

The issue of co-infestation with multiple parasites is also relevant to the influence of lice and other ectoparasites on host mate choice and sexual selection. For example, bill color in common blackbirds (*Turdus merula*) is thought to be a sexually selected trait. Biard et al. (2010) found no correlation between the bill color of common blackbirds (*Turdus merula*) and the abundance of each of several parasite taxa. However, there was a strong correlation between bill color and combined measures of parasite load, including lice.

Costs of host resistance

In the next chapter we review adaptations that mammals and birds have evolved for resisting lice. Here we preview the physiological and other costs of mounting such resistance because these costs represent indirect effects of lice on host fitness. For example, grooming—which includes oral grooming in mammals and preening in birds—are both effective defenses against lice. Most species of birds spend 5–15% of their time preening (Cotgreave and Clayton 1994). Preening is responsive to increases in louse load (Hart 1997), suggesting that time spent preening is a stimulus-driven inducible defense, rather than a programmed constitutive defense (Mooring et al. 2004). Birds parasitized by more species of lice devote more time to preening (Cotgreave and Clayton 1994).

Preening is expensive in terms of both time and energy (Goldstein 1988). Croll and McLaren (1993) documented nearly a 200% increase in the metabolic rates of preening thick-billed murres (*Uria lomvia*), compared to resting individuals. Notably, this increase was higher than increases in metabolic rate associated with either feeding (49%) or diving (140%). Breeding king penguins (*Aptenodytes patagonicus*) spend about 22% of their time in "comfort" behavior, 16% of which is preening. Penguins engaged in comfort behavior have metabolic rates that are about 24% higher than the resting metabolic rate. This increase in energy expenditure is presumably risky because king penguins must fast for weeks at a time while incubating the eggs (Viblanc et al. 2011). If energy reserves get too low, the penguins will abandon their eggs.

Grooming in mammals can also be energetically expensive, as suggested by anecdotes indicating that poorly nourished individuals do not groom as much as well-fed individuals (Murray 1990). The energetic cost of grooming to control mammal lice has not been measured, to our knowledge. However, since mammals with more lice spend more time grooming (Mooring et al. 1996), they presumably expend more energy.

In contrast, the energetic cost of grooming to control parasitic bat mites has been measured (Giorgi et al. 2001). The grooming times and metabolic rates of mouse-eared bats (*Myotis myotis*) experimentally infested with blood-feeding mites are shown in fig. 3.5. Bats with 40 mites nearly tripled their grooming time, compared to controls without mites. Bats with 40 mites also increased their metabolic rates by 21.3%, compared to controls without mites. Thus, grooming to control mites is energetically costly, at least in bats.

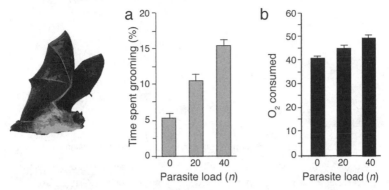

FIGURE 3.5. Effect of mites on the grooming and energetics of mouse-eared bats (*Myotis myotis*). (a) Mean (+se) time spent grooming, and (b) mean (+se) oxygen consumption (mL O_2 h^{-1}) for 14 bats experimentally infested with the numbers of mites shown. (a) Grooming time was correlated with mite load (P < 0.001). (b) Oxygen consumption was also correlated with mite load (P < 0.001). After Giorgi et al. (2001).

Increased grooming can also involve other costs, such as reduced vigilance against predators, which has been shown for both birds (Redpath 1988) and mammals (Cords 1995; Mooring and Hart 1995). Reduced vigilance can lead to an increase in attacks on offspring by conspecifics (Maestripieri 1993). Increased grooming may also lead to reduced vigilance against rival males for oestrous females (Hart et al. 1992). Yet another cost associated with grooming in mammals is the loss of water in saliva generated by grooming. In rats, this has been shown to account for up to one-third of the animal's daily water budget (Ritter and Epstein 1974). Grooming creates wear and tear, such as the attrition of dental elements used by ungulates when they groom (see chapter 5) (McKenzie 1990). Finally, grooming can lead to winter hair loss and thermoregulatory costs, as shown in moose (*Alces alces*) infested with ticks (Mooring and Samuel 1999).

Data on the costs of oral grooming and preening should be interpreted with caution because these behaviors serve a variety of roles in addition to parasite control—for example, they are also important for straightening and cleaning pelage and plumage. Moreover, grooming plays important roles in defense against ectoparasites other than lice, such as fleas, mites, and ticks. It would be interesting to know the relative cost of grooming for controlling these different groups of parasites on mammals and birds. Experimental infestations with different parasite taxa, individually and in combination, could be informative.

Costs of defense also apply to sexually selected traits evolved by males to advertise freedom from lice, as discussed above. The physiological costs as-

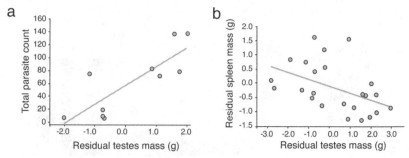

FIGURE 3.6. (a) Relationship between number of sucking lice and residual testis mass (RTM) (corrected for body size) in the South African ground squirrel (*Xerus inauris*). The relationship was highly statistically significant (P = 0.003), with RTM explaining 68% of variation in louse intensity; (b) RTM was also significantly negatively correlated with residual spleen mass (P = 0.02); RTM explained 21% of variation in this variable. After Manjerovic and Waterman (2012).

sociated with ornamental plumage, display behavior, and male-male combat help to keep signals "honest" (Andersson 1994). In addition to the costs of producing and exhibiting such traits, there may be costs associated with maintaining them in good condition. Walther and Clayton (2005) tested this "high maintenance" handicap principle and showed that bird species with more ornamental plumage devote significantly more time to maintenance behavior than sister species with less ornamental plumage (plate 6). Thus, ornamental plumage can be costly simply because of the time and energy required to keep it in good condition.

Immunological defenses are also costly (Sheldon and Verhulst 1996; Schmid-Hempel 2011). A large body of work has been devoted to negative trade-offs between the costs of immunological defense and host reproductive success, including the expression of parasite-indicative traits (Folstad and Karter 1992; Poiani et al. 2000). Unfortunately, studies involving bird lice have relied mainly on feather hole data (box 3.1). The mammal data are more convincing. Manjerovic and Waterman (2012) documented a strong positive correlation between testes size and the number of lice on South African ground squirrels (*Xerus inauris*), a promiscuous species with one of the highest male reproductive investments known among rodents (fig. 3.6a). The authors found that testes size is negatively correlated with spleen mass (fig. 3.6b), which they used as a measure of immune strength (Corbin et al. 2008). Thus, male ground squirrels investing more in reproduction may have poorer immune defenses, resulting in more lice.

Host tolerance

Hosts may avoid the costs of defense if they are able to tolerate small populations of lice without suffering a reduction in fitness. Differently put, tolerance is the ability to minimize the negative effect of parasites on fitness without reducing the number of parasites. Long known to plant biologists (Caldwell et al. 1958), tolerance has been overlooked by most animal biologists until recently (Blanchet et al. 2010; Svensson and Råberg 2010; Medzhitov et al. 2012). The topic is confusing because the term "tolerance" is also used in reference to immunological tolerance, a different phenomenon in which B- and T-cells stop responding to self through the elimination, conversion, or inactivation of self-reactive lymphocytes during maturation (Schmid-Hempel 2011).

Increases in tolerance do not select for counter-adaptations in parasites because, unlike host resistance, tolerance does not have a negative effect on parasite fitness (Svensson and Råberg 2010). Trade-offs between tolerance and resistance exist and may be common (Råberg et al. 2007). Experimental work indicates that laboratory and domesticated animals do vary in their tolerance to ectoparasites (Nelson et al. 1977). For example, Stewart et al. (1976) reported differences in the ability of grooming-impaired lab mice to tolerate experimental infestations with the sucking louse *Polyplax serrata*. The authors conducted experiments to determine whether early exposure to lice leads to greater tolerance, but this was not the case. Wild birds vary in their tolerance to fleas (Heeb et al. 1998) and botflies (Norris et al. 2010). Variation in tolerance to lice has not been studied in wild birds or mammals to our knowledge.

Wild mammals and birds are often chronically infested with small numbers of lice (Durden 2001; Clayton et al. 2008). Selection may not favor the evolution of resistance mechanisms that eliminate these lice, given the costs of resistance discussed above. Instead, selection may favor mechanisms that simply limit parasites to population sizes hosts can tolerate. The damage done by small populations of lice is probably minimal (unless the lice are vectoring dangerous pathogens or other parasites).

It would be interesting to test whether birds vary in their tolerance to feather damage by feather-feeding lice. As discussed earlier in this chapter, large numbers of lice can create enough damage to have severe effects on host fitness. However, birds may tolerate small numbers of feather-feeding lice. The mechanism responsible could be as simple as evolving a slight increase in feather number, which varies within bird species (Wetmore 1936). It would be interesting to test whether individual birds with more feathers are more tolerant to feather-feeding lice, and whether louse-mediated selection favors the evolution of "extra" feathers.

II COADAPTATION

Our first attempts to seize what Japanese macaques pick up
during grooming resulted in failure, since the adult monkeys
grimaced and threatened us.
—Tanaka and Takefushi 1993

I n the last chapter we reviewed ways in which lice are known
to reduce the fitness of their hosts. Birds and mammals have
evolved a variety of adaptations for resisting lice and reducing
their effects on host fitness. Some of these adaptations appear
specific to lice, while others are also effective against other ectoparasites. Overall, birds and mammals have three broad strategies for deal-
ing with lice: 1) avoidance, 2) tolerance, and 3) resistance. Although individual hosts cannot choose their parents, they can choose a louse-free mate,
which is an example of the first strategy. Parasite-mediated mate choice
was reviewed in the last chapter. We also considered the second strategy—
tolerance of small infestations—at the conclusion of the last chapter. In
this chapter we review adaptations for resisting lice, which is the third and
most common defense strategy. Resistance adaptations include molting of
plumage or pelage, endogenous chemical resistance, maintenance behavior, and immunological responses. We conclude the chapter by considering
the genetics of resistance.

Molting of plumage or pelage

Lice depend on plumage or pelage (hair or fur) for habitat. One way to
combat lice is to jettison this habitat by molting or shedding it, not unlike
plants dropping their leaves to combat leaf-mining insects (Stiling et al.
1991). Murray (1957) documented an 80% reduction in lice eggs on shedding
horses. Baum (1968) reported an 85% drop in the abundance of lice on molting Eurasian blackbirds (*Turdus merula*). In contrast, Moyer et al. (2002b)
found that molt did not reduce populations of lice on rock pigeons, partly
because the lice used the newly emerging pin feathers as refuges from molt.
Thus, although molt may combat lice in some cases, this is not always true.

It has long been suggested that the evolutionary loss of thick body hair
by humans may have been an adaptation for combating lice and other ectoparasites (Belt 1874; Darwin 1888; Rantala 1999; Pagel and Bodmer 2003;
Dawkins 2005). This hypothesis suggests that the short, fine body hair of
humans reduces the amount of suitable habitat for ectoparasites. Loss of

body hair was made possible when early hominids invented clothing and fire, eliminating the need for thick hair to stay warm. Interestingly, human body lice have evolved the ability to use clothing as habitat instead of hair. This, in turn, helps date the cultural evolution of clothing use by humans (see box 9.2).

Endogenous chemical resistance

Some birds and mammals produce defensive chemicals for combating lice and other ectoparasites, analogous to the chemical defenses of plants against herbivorous insects (Painter 1958; Ehrlich and Raven 1964; Strong et al. 1984; Fritz and Simms 1992). The most thoroughly studied and discussed example of chemical defense against ectoparasites involves batrachotoxins in the skin and feathers of several species of New Guinea birds in the genera *Pitohui* and *Ifrita* (Dumbacher et al. 1992, 2000; Dumbacher and Pruett-Jones 1996; Mouritsen and Madsen 1994; Poulsen 1994). Dumbacher (1999) conducted a series of trials in which he exposed feather lice in petri dishes to feathers of *Pitohui* species, and feathers from other (nontoxic) birds. Lice avoided the *Pitohui* feathers when given a choice, and lice constrained to *Pitohui* feathers showed higher mortality than lice on nontoxic feathers.

It is well known that different sexes, breeds, and species of mammals vary in their attractiveness to blood-feeding flies and ticks (Weldon and Carroll 2006). Chemicals produced by some bird and mammal species may also repel lice. At least 80 genera of birds in 17 orders have pungent odors that are easily detected by humans (Hagelin and Jones 2007). For example, crested auklets (*Aethia cristatella*) secrete aldehydes with a citrus-like odor that can be smelled by humans more than a kilometer away (Jones 1993). These aldehydes include hexanal and octanal, which are known arthropod repellents. Douglas (2013) showed that exposure to auklet aldehydes, or synthetic versions of them, can paralyze or even kill lice. These results contrast with earlier work suggesting that auklet odors have little or no effect on lice (Douglas et al. 2005). This interesting topic requires further work.

The preen oil that birds secrete and spread through their plumage may also combat lice and other ectoparasites. However, data relevant to this hypothesis are conflicting (box 4.1).

Another compound with the potential to combat lice is the pigment melanin, which is involved in brown, grey, and black feathers of birds (McGraw 2006). Melanin is also known to increase the resistance of feathers to mechanical abrasion (Burtt 1986; Bonser 1995). A study of barn swallows (*Hirundo rustica*) suggested that melanin deters feather-feeding lice (Kose et al. 1999). However, experiments by Bush et al. (2006a) found that melanin

BOX 4.1. Do birds use preen oil to combat lice?

Uropygial gland secretions, or "preen oil," which birds distribute throughout their feathers when preening, may be important for defense against ectoparasites (Clayton et al. 2010) and feather-degrading bacteria (Soler et al. 2012; Czirják et al. 2013). To test the hypothesis that preen oil combats lice, Moyer et al. (2003) compared the survival of lice raised in an incubator on feathers treated with uropygial secretions to the survival of lice on control feathers without secretions. The authors found that lice on feathers with secretions died significantly faster than lice on feathers without secretions. However, mineral oil had the same effect, which was apparently to block the spiracles (breathing pores) of the lice.

Moyer et al. (2003) also compared populations of lice on captive pigeons from which glands were surgically removed to populations of lice on control pigeons from which the glands were not removed. There was no difference in the number of lice on birds in the two groups over a four-month study, suggesting that uropygial secretions do not, in fact, combat pigeon lice.

Moreno-Rueda (2010, 2014) reported negative correlations between uropygial gland size and the number of holes in the primary and secondary wing feathers of house sparrows (*Passer domesticus*). On this basis, he suggested that uropygial secretions combat the lice that are thought to make holes in feathers (but see box 3.1).

Møller et al. (2010) reported a correlation between uropygial gland mass (corrected for overall body mass) and the generic richness of amblyceran lice (but not ischnoceran lice) across 212 species of birds. The authors suggested that Amblycera may have diversified more than Ischnocera because Amblycera feed on blood and "had to cope with two defense systems: immunity and wax from the uropygial gland." This hypothesis seems premature, pending a demonstration that uropygial gland secretions are, in fact, detrimental to lice.

In summary, the evidence for any defensive function of preen oil against lice remains weak.

does not affect the feather-feeding lice of rock pigeons. The authors captured feral pigeons that varied in color from white to dark grey. Lice were fed feathers from these birds in vitro over a period of two weeks, but there was no significant difference in the quantity of feather material consumed, nor in the survival of lice on the different colored feathers. Further experiments revealed no difference in the reproductive success of pigeon lice on white vs. black feathers, nor did lice exhibit a preference for feathers of different colors.

In summary, there is evidence that a few species of birds produce chemicals that repel, disable, or even kill lice. However, chemical defense probably plays little or no role in combating the lice of most birds or mammals. It certainly does not compare to the importance of chemical defenses of plants against herbivorous insects (Núñez-Farfán et al. 2007). One reason for the difference is that birds and mammals have complex anti-parasite behavior. Parasite-mediated selection has adapted some of these behaviors into very effective mechanisms for resisting lice and other ectoparasites, as we review in the next section.

Behavioral resistance

The first, and arguably most effective, line of defense against mammal and bird lice is grooming and other "maintenance" behaviors. Mammals use their teeth or tongues for oral grooming, while birds use their beaks for preening. Many mammals also rely on mutual grooming, or "allogrooming," to combat ectoparasites on regions that cannot be self-groomed, such as the head and neck (plate 7). Mammals and birds also groom by scratching with their feet to control lice on the head and upper body. Both host groups have a variety of morphological adaptations for grooming. They also have other kinds of maintenance behaviors, such as rubbing, wallowing, dusting, and sunning, that appear to help resist lice and other ectoparasites. These defenses are reviewed below.

Oral grooming in mammals

Oral grooming is a very efficient means of controlling lice in mammals, such as rodents, felids, ruminants, and primates (Mooring et al. 1996). When mice (*Mus domesticus*) infested with lice (*Polyplax serrata*) were given collars to impair their grooming ability, the mean number of lice increased from less than 100 to more than 2,000 in three weeks. Upon removal of the collars, the lice returned to normal levels within a day (Murray 1961, 1990). Tracking the distribution of lice on the body of the host is another way to illustrate the efficiency of grooming. Lice are normally found on the head and neck of mice, but when grooming is prevented, lice spread across the rest of the body. If grooming is prevented on one side of the body, lice spread across that side of the body only (Murray 1961, 1987). The importance of grooming is also reflected in the amount of time mammals devote to it. Rats spend 30–50% of their time grooming, making it the most time-consuming portion of their daily time budget, other than sleeping (Bolles 1960).

Host body size has a fundamental effect on grooming patterns. Small-bodied species spend more time grooming than large-bodied species, compensating for the relatively higher cost of ectoparasite damage on bodies with greater surface to volume ratios (Mooring et al. 2000). Large mammals spend less time grooming, and they are not as thorough when they groom. Although grooming by larger-bodied hosts is less thorough, it can still be effective in controlling the overall abundance of lice. When cattle infested with sucking lice (*Linognathus vituli*) and chewing lice (*Bovicola bovis*) were restrained from grooming, their lice increased by 15-fold (Lewis et al. 1967). Extremely large mammals, such as African elephants (*Loxodonta africana*) and giraffes (*Giraffa camelopardalis*), spend little or no time grooming. Instead, they appear to rely on wallowing, dusting, and other behaviors to combat lice and other ectoparasites (Mooring et al. 2000).

Some mammals have morphological adaptations for grooming. For example, mice pull their fur through lower incisors that can move laterally. If lateral movement is prevented by experimentally propping the teeth apart, louse populations increase rapidly, even when the mice continue normal amounts of grooming (Murray 1961). Cattle groom with their teeth, as well as their tongues, which act like coarse brushes. Oxen have tongues covered with papillae, reminiscent of the small teeth on currycombs (round brushes used to groom horses) (Murray 1987).

Other ungulates, such as antelope and deer, use the lateral edges of their lower incisors and canines to groom with an upward sweeping motion (Hart et al. 1992; McKenzie 1990). In impala (*Aepyceros melampus*) and other medium-sized antelopes, these teeth make up the "lateral dental grooming apparatus" (fig. 4.1a). As in the case of mouse incisors, the teeth move laterally, which assists in parasite control (McKenzie and Weber 1993). Impala with experimentally impaired grooming have many more ticks than control animals (Mooring 1995). Grooming presumably also helps to control the five species of lice that impala are known to harbor (Matthee et al. 1998). Dental modifications used in grooming have evolved in other unrelated groups of mammals. For example, prosimian primates such as lorises, lemurs, and bushbabies all have mandibular toothcombs (fig. 4.1b; Rose et al. 1981). Toothcombs are also present in hyraxes, treeshrews, and cologos (flying lemurs).

Allogrooming in mammals
Allogrooming of one individual by another also plays a central role in the control of lice on mammals (Murray 1987). Mice are capable of controlling lice by allogrooming even if self-grooming is impaired. However, effective allogrooming depends on a stable social hierarchy; when this hierarchy is disturbed, populations of lice increase rapidly (Lodmell et al. 1970). Allogrooming is often directed at regions that cannot be self-groomed, such as the head. Allogrooming in larger mammals, such as impala, is directed

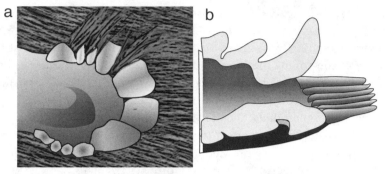

FIGURE 4.1. (a) Lateral dental grooming apparatus of an impala moving through the pelage (drawn by S.E.B. from a photograph of simulated grooming behavior in MacKenzie 1990); (b) Mandibular toothcomb formed by six incisors of a needle-clawed bushbaby (*Euoticus elegantulus*) (drawn by S.E.B. from a photograph in Rose and Walker 1981).

more broadly over the surface of the body (Hart et al. 1992). In some species, allogrooming takes place mainly between the members of one sex. For example, in Cape ground squirrels (*Xerus inauris*) allogrooming is mainly between females, which therefore have smaller ectoparasite loads than males (Hillegass et al. 2008).

Primates use their fingers, as well as their teeth, to remove lice and other ectoparasites by allogrooming (Tanaka and Takefushi 1993). Many of the lice are eaten, but it is not known if this has significant nutritional benefit (Johnson et al. 2010). It is conceivable that consumed lice may provide trace elements or vitamins. Primates spend up to 18% of their total time allogrooming (Grueter et al. 2012). The efficiency of allogrooming has been quantified in Japanese macaques (*Macaca fuscata*) (plate 7a; Tanaka and Takefushi 1993), which must locate and remove 50–100 louse eggs per day to keep their populations in check (Zamma 2002a). Allogrooming is directed at body regions with the greatest density of lice, rather than the greatest surface area.

Components of grooming in primates can be complex and culturally inherited. Tanaka (1995, 1998) documented the trial and error discovery by a female macaque of increasingly efficient methods of grooming. The macaque's sister, daughters, and granddaughter subsequently learned to use the same methods after observing her. Fig. 4.2 shows the relative efficiency, over time, of different methods used by one of the daughters.

Another example of complex behavior is the use of leaves by common chimpanzees (*Pan troglodytes*) as "napkins" for inspecting and crushing ectoparasites removed from other chimps by allogrooming (Assersohn et al. 2004). Zamma (2002b) videotaped a chimpanzee using a leaf to in-

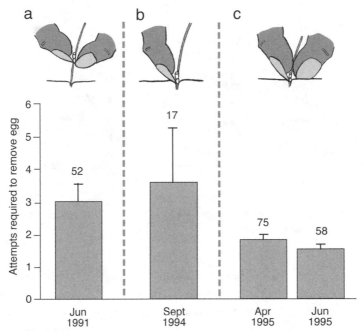

FIGURE 4.2. Relative efficiency of grooming techniques for removing 0.5-mm louse eggs. The eggs are attached to hairs with rings of adhesive cement. (a) One-hair combing technique using the slit between two nails (the egg is not loosened first); (b) Forefinger nail technique used to loosen the egg; (c) Forefinger nail loosening with the thumb used as a jig. Number of bouts observed above each bar. After Tanaka (1998).

spect and crush a louse that it had removed from another chimp that it was allogrooming with its teeth (plate 8). Similarly, Zamma (2006) witnessed a chimp using a leaf as a tool to manipulate a louse egg it had just removed from another chimp (fig. 4.3).

Given these observations, it is interesting to speculate whether anti-parasite behavior may have contributed to the evolution of dexterous tool use in hominids. In a related vein, anthropologists have argued that human language evolved as a form of "verbal grooming" facilitating social cohesiveness (Dunbar 1993, 1996, 2003; Huron 2001). The hypothesis is that, over time, verbal grooming may have replaced physical grooming as the size and mobility of hominid groups increased beyond the level at which physical grooming could provide sufficient social "glue" (but see Grueter et al. 2012).

Homo sapiens has long relied on grooming, or "nitpicking," for the control of lice. Ancient nit combs—some with long-dead nits attached—have been recovered from archaeological sites around the world (Mumcuoglu 2008a,b). Human allogrooming is still practiced around the world (plate 7b),

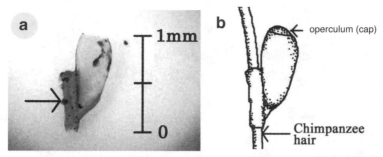

FIGURE 4.3. (a) Photograph of a hatched egg recovered from a leaf used as a "napkin" by a chimpanzee following a bout of oral allogrooming. The ring of adhesive cement (arrow) was unbroken, indicating that the egg was removed by pulling it along the length of the hair to which it was attached; (b) Interpretive drawing of unhatched egg attached to a hair. Both images from Zamma (2006).

and in many cultures the lice are eaten (Nuttall 1917; Zinsser 1935; Murray 1990; Overstreet 2003). In other cultures, nitpicking is viewed with disgust, yet practiced out of necessity (Roberts 2002; Burgess 2004). Allogrooming in humans and other mammals also facilitates pair-bonding (Nelson and Geher 2007). Curtis and Biran (2001) suggest that the strong emotional aversion to lice and other parasites is a behavioral adaptation to avoid infection. However, the role of lice as vectors may have influenced other forms of social behavior in humans (chapter 2).

Preening in birds

Preening is the first line of resistance for combating lice on birds (plate 9a). Brown (1972) found that chickens with large numbers of lice preen more than controls without lice. Similarly, Villa et al. (unpublished data) showed that pigeons with lice preen significantly more than pigeons without lice. Analysis of the stomach contents of birds suggests that, unlike primates, birds do not eat many of the lice they remove (Ash 1960). Experimental work confirms that rock pigeons consume less than 5% of the lice removed during preening (Bush 2004).

The important role of preening in the control of bird lice was first suggested by "natural experiments" in which birds with deformed bills were observed to often have very large numbers of lice (Ash 1960; Pomeroy 1962; Clayton 1991a; Clayton et al. 1999). However, birds with deformed mandibles may have other problems, such as impaired foraging ability that results in poor general condition. Clayton (1991a) tested the effect of preening on lice by fitting rock pigeons with poultry "bits," which are small C-shaped pieces of metal or plastic (fig. 4.4). Bits create a gap of 1–3 mm between

FIGURE 4.4. Rock pigeon fitted with a poultry bit, which creates a gap of 1–3 mm between the mandibles of the bill that impairs the bird's ability to control lice by preening. Photograph by S.E.B.

the mandibles that impairs the occlusion of the bill required for efficient preening. Bits trigger dramatic increases in the feather louse populations on pigeons (Clayton 1990b, 1991a; Clayton and Tompkins 1995; Clayton et al. 1999). These increases are not due to side effects of bits, which do not interfere with feeding in pigeons. Clayton and Tompkins (1995) showed that bits have no effect on the survival or reproductive success of (louse free) rock pigeons, compared to controls without bits.

Among birds of the world, beaks differ dramatically in size and shape (fig. 4.5). If birds with long, unwieldy beaks are less efficient at preening, then they may have more lice than birds with short beaks. Clayton and Walther (2001) tested this prediction for 34 species of Peruvian birds with diverse beak shapes and sizes; however, there was no relationship between beak size and louse abundance. The same study pointed to an unexpected negative relationship between louse abundance and the length of the over-hanging tip of the upper mandible of the beak (fig. 4.6a). The negative correlation suggested that birds with longer overhangs are better at controlling lice by preening than birds with short overhangs. (Extreme overhangs, like the hooked beaks of raptors and parrots, are clear adaptations for feeding, not preening.) Freed et al. (2008b) also reported a negative relationship between beak overhang length and the prevalence and intensity of lice among eight species of Hawaiian birds, including six species of native honeycreepers (Drepanidinae).

Clayton et al. (2005) tested the role of the beak overhang by trimming it from wild-caught rock pigeons (fig. 4.6a,b). This procedure caused a dramatic increase in lice on the trimmed birds (fig. 4.6c). When the overhangs were allowed to grow back, birds regained their ability to control their lice by preening. Trimmed and non-trimmed birds did not differ in the amount of time they spent preening, so the effect on lice was caused by the overhang itself, not compensatory preening (Clayton et al. 2005).

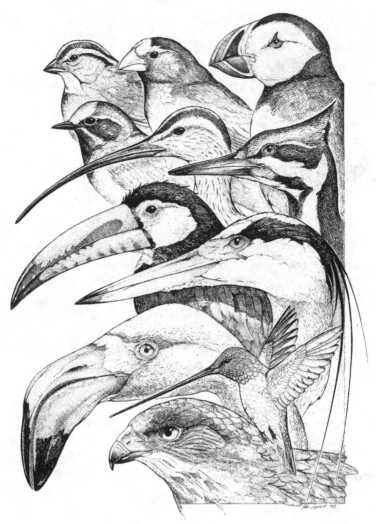

FIGURE 4.5. Birds of the world have an impressive diversity of bill sizes and shapes. From Proctor and Lynch (1993), with permission of Yale University Press.

High-speed video revealed that, when pigeons preen, the lower mandible moves forward against the beak overhang (fig. 4.6d, e). This motion, which is very fast—at up to 32 times per second—creates a shearing force that cannot be generated in the absence of the overhang. In summary, the beak overhang functions as a template against which the tip of the lower mandible exerts pressure when preening. The force is sufficient to severely damage or kill lice, despite their tough exoskeletons and dorsoventrally compressed shape (plate 10).

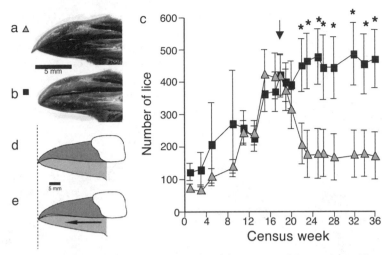

FIGURE 4.6. Rock pigeon beak before (a) and after (b) trimming of the overhang; (c) mean (± se) abundance of lice on 26 adult pigeons. The overhangs of all birds were trimmed for 17 weeks. At week 18 (arrow) overhangs on half the birds were allowed to grow back (grey triangles), while remaining birds continued weekly trimming (black squares). There were significant overall effects of trimming and time (P < 0.01), and a significant interaction between treatment and time (P < 0.0001). Birds with overhangs had significantly fewer lice than trimmed birds at each of the final eight censuses. (d, e) Beak tracings from high-speed video of a rock pigeon preening a neck feather; the tracings are before (d), and after (e) the forward movement of the lower mandible. After Clayton et al. (2005).

One might predict that the overhang also plays a role in feeding. To test this hypothesis, Clayton et al. (2005) compared the time required for hungry pigeons to pick up seeds of different sizes before and after manipulation of the beak. Remarkably, however, there was no effect of removal of the overhang on foraging efficiency for any of three seed types. The overhang was critical for controlling lice, but played no role in feeding efficiency.

Although these experiments confirm that the overhang is an essential feature of preening for louse control, at least in pigeons, they do not explain why the overhang is a mean of 1.5 mm in length, not longer. One potential cost of longer overhangs is that they could be more susceptible to damage. To test this hypothesis, Clayton et al. (2005) examined the risk of damage to the overhang in relation to length. Their results showed that longer overhangs do, in fact, break significantly more often than short overhangs (fig. 4.7). Since birds with broken overhangs cannot preen efficiently, lice presumably exert stabilizing selection for overhangs of intermediate length.

The stabilizing selection hypothesis has also been tested in Western scrub-jays (*Aphelocoma californica*), which have variable beak overhangs

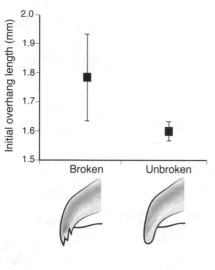

FIGURE 4.7. Long overhangs break significantly more (P < 0.05). The overhangs of 13 rock pigeons that suffered breaks were significantly longer at the start of the month-long experiment than those of 111 birds without breaks. After Clayton et al. (2005).

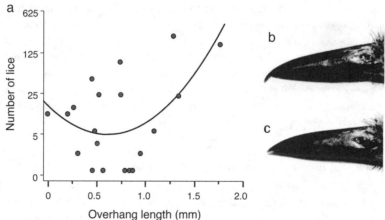

FIGURE 4.8. (a) Abundance of lice on western scrub jays (*Aphelocoma californica*) with variable beak overhang lengths (note log scale on y-axis). Birds with intermediate overhangs had significantly fewer lice than birds with longer (b) or shorter (c) overhangs, consistent with a pattern of stabilizing selection on overhang length (quadratic regression $R^2 = 0.30$, $P < 0.05$). After Clayton et al. (2010).

(Peterson 1993; Moyer et al. 2002c). Jays with intermediate overhangs have significantly fewer lice than jays with overhangs that are either shorter or longer (fig. 4.8). Although no test of the effect of lice on scrub-jay fitness has been carried out, Hoi et al. (2012) recently reported significant effects of experimentally increased feather lice on the fitness of European bee-eaters (*Merops apiaster*), which are similar in size to scrub-jays and have related lice (fig. 3.2).

In contrast to the results for lice, experimental removal of the beak over-hang has no effect on the ability of pigeons to kill hippoboscid flies, the other common group of pigeon ectoparasites controlled by preening (Waite et al. 2012). In summary, the data from these experiments suggest that the beak overhang is a specific adaptation for resisting lice, at least in pigeons. The influence of beak morphology on preening shows that some features of the beak have evolved in response to parasite-mediated selection, rather than foraging-mediated selection. It would be interesting to explore trade-offs between the functional morphology of beaks for preening versus for-aging among birds of the world.

Allopreening in birds

Mutual preening, or allopreening (plate 9b), may help to control bird lice, just as allogrooming helps to control mammal lice. Unfortunately, much less is known about allopreening. It is known to be important for controlling ticks on the heads and necks of Macaroni penguins (*Eudyptes chrysolophus*) (Brooke 1985). Similarly, allopreening in green woodhoopoes (*Phoeniculus purpureus*) is thought to control ectoparasites, including lice (Radford and Plessis 2006). Unfortunately, however, no rigorous tests of the effectiveness of allopreening in controlling bird lice have been carried out (Clayton et al. 2010). One test might involve the analysis of covariation between allopreening rates and parasite load (cf. Mooring 1995). However, an experimental test of the effectiveness of allopreening in combating lice is also needed.

Scratching in mammals and birds

Like allogrooming, foot scratching helps to control lice on the head and neck. The effectiveness of scratching is shown by sharp increases in lice on the foreparts of mice missing their hind toes (Bell et al. 1962; Ratzlaff and Wikel 1990). Primates, hyraxes, and some other mammals have modified "grooming claws," or "toilet claws," that are used for grooming (Maiolino et al. 2011). Unfortunately, the effectiveness of these interesting struc-tures for resisting lice or other ectoparasites has not been tested to our knowledge.

Birds with a deformed foot often have large numbers of lice and eggs restricted to the head and neck, which they cannot preen, nor scratch while standing on the good foot (Clayton 1991a). Although the proximal effect on lice has not been measured, scratching is known to damage and kill chicken fleas, which are also heavily sclerotized (Marshall 1981a). Scratching may also compensate for inefficient preening in birds with unwieldy bills. Long-billed species average 16.2% of their grooming time scratching, compared

FIGURE 4.9. (a) Barn owl (*Tyto alba*) scratching; (b) pectinate claw from the middle toe of a barn owl. Photo (a) by Mike Read, naturepl.com, and (b) from Bush et al. (2012).

to just 2.3% of grooming time in short-billed species. Moreover, in comparisons of closely related species, long-billed species scratch significantly more often than short-billed species (Clayton and Cotgreave 1994). Clayton and Walther (2001) tested for a relationship between foot morphology and the richness and abundance of lice in 34 species of Peruvian birds; however, they found no significant correlations.

Some birds have pectinate claws that may improve the ectoparasite control function of scratching. In a survey of 118 families of birds, Clayton et al. (2010) found pectinate claws in 17 families representing 10 orders. For example, barn owls (*Tyto alba*) have pectinate claws with 5–12 clearly defined teeth (fig. 4.9). Bush et al. (2012) found that owls having claws with more teeth were less likely to be infested with lice than owls having claws with fewer teeth. However, an experimental manipulation (cf. fig. 4.6) is needed to confirm the function of pectinate claws for combating lice and other ectoparasites.

Dusting and other substrate grooming

Mammals engage in dust-tossing, mud-bathing, wallowing, and rubbing behaviors (plate 11a; Mooring et al. 2000). These behaviors are thought to help combat lice. For example, wallowing is thought to reduce lice on Asian water buffalos (*Bubalus bubalis*) (Mitzmain 1912; Weilgama 1999). However, no test of the effectiveness of any of these behaviors has been carried out to our knowledge.

Members of at least a dozen orders of birds also engage in dusting, during which fine dirt or sand is ruffled through the feathers, accompanied by preening (plate 11b). This behavior is thought to remove excess feather oil that can matt the plumage (Healy and Thomas 1973; Borchelt and Duncan 1974; Van Liere 1992). It also appears to combat lice through desiccation, which kills lice outright, or makes them more vulnerable to preening. Martin and Mullens (2012) allowed chickens with lice to dust using sand, litter, or kaolin (fine clay). Birds dusting with kaolin showed dramatic reductions in their lice. However, dusting with sand or litter had no appreciable effect on lice.

Anointing behavior

Another intriguing possible mechanism for combating lice and other ectoparasites is anointing behavior, in which animals "apply scent-laden materials to their integument" (Weldon and Carroll 2006). Mammals and birds anoint with a remarkable variety of substances, including millipedes, caterpillars, ants, beetles, snails, toads, carrion, resin, citrus (fig. 4.10), onions, walnut juice, flowers, leaves, smoking vegetation, mothballs, lawn chemicals, and the urine, feces, or skin gland secretions of other species. Unfortunately, the effectiveness of most of these substances in combating lice or other ectoparasites is unknown (Clayton et al. 2010; Alfaro et al. 2011; Weldon et al. 2011).

Clayton and Vernon (1993) tested whether citrus kills lice. After observing a common grackle (*Quiscalus quiscula*) anointing its feathers with a lime, the authors measured the effect of lime on pigeon lice in vitro. Lime juice had no effect, but exposure to vapor from lime rind was lethal. The rind

FIGURE 4.10. Captive boat-tailed grackle (*Quiscalus major*) anointing with a slice of lemon. Vapor from citrus rind has been shown to kill lice, in vitro. Images from video; courtesy of Paul Weldon and Jacquie Greff, Tonal Vision.

contains D-limonene, a monoterpene that is toxic to guinea pig lice (Hink and Fee 1986). Citronella and other citrus components are also known to repel lice (Mumcuoglu et al. 2004) and other ectoparasites (Weldon et al. 2011).

Plant products are also used by humans to combat lice. For example, pyrethrins are naturally occurring insecticides from chrysanthemum flowers that have been used to control lice on humans and domesticated animals since the 1940s (Burgess 2004). Other plant products, such as essential oils, are capable of repelling or even killing lice (Khater et al. 2009; Priestley et al. 2006; Rossini et al. 2008; Talbert and Wall 2012).

One of the most intriguing forms of anointing behavior is "anting," in which birds crush and smear ants into their plumage (active anting), or lie on ant mounds or trails and allow ants to crawl through their feathers (passive anting) (Whitaker 1957; Simmons 1966). Anting has been observed in over 200 species of birds, mostly passerines. The fact that birds ant exclusively with ants that secrete formic acid or other pungent fluids suggests the behavior may kill or deter lice or other ectoparasites. Unfortunately, there is no compelling evidence in support of this hypothesis (Clayton et al. 2010).

Primates have also been observed anting (Longino 1984; Alfaro et al. 2011), which could repel mosquitoes or ticks (Falótico et al. 2007). However, as in the case of birds, there appears to be no conclusive evidence that anting controls lice in primates, although some workers are convinced that it does serve this function (Verderane et al. 2007). Nuttall's (1917) anecdote that British and German soldiers with body lice "reduced their louse population by placing their shirts on ant hills" is also intriguing.

Nest fumigation

Birds are also known to resist lice and other parasites by fumigating their nests with aromatic vegetation (Weldon and Carroll 2006). European starlings (*Sturnus vulgaris*) incorporate species of plants containing chemicals with antibacterial and insecticidal properties into their nests, leading to reductions in the hatching success of lice (*Menacanthus* sp.) (Clark and Mason 1985). Wild carrot (*Daucus carota*) or fleabane (*Erigeron philadelphicus*) added to nests reduces the emergence of ectoparasitic mites (*Ornithonyssus sylviarum*) (Clark and Mason 1988). Indeed, the name "fleabane" is derived from the fact that its flowers are said to repel and kill fleas and other insects in households (Panagiotakopulu et al. 1995).

Another recent study (Suárez-Rodríguez et al. 2013) showed that urban house sparrows (*Passer domesticus*) and house finches (*Carpodacus mexicanus*) weave cellulose fibers from smoked cigarette butts into their nests, and that nests with more fibers have significantly fewer mites. The authors also showed that mites are repelled by fibers from smoked cigarettes, presumably because smoking releases nicotine, which has been used to control arthropod pests on agricultural crops (Rodgman and Perfetti 2008) and poultry (Lans and Turner 2011). Whether incorporating cigarettes into nests would help combat lice is unknown.

Mammals also fumigate their nests or burrows with vegetation. For example, nests of dusky-footed woodrats (*Neotoma fuscipes*) sometimes contain bay leaves (*Umbellularia californica*), which kill cat fleas under laboratory conditions (Hemmes et al. 2002). These results are consistent with the hypothesis that woodrats fumigate their nests to control ectoparasites.

Sunning behavior

Species in at least 50 families of birds are known to adopt stereotyped postures and sun themselves (plate 11c) (Simmons 1986). Sunning birds often pant or show other signs of heat stress. Sunning has interesting parallels with "behavioral fever," in which ectothermic animals kill pathogens and other parasites by choose to bask in warm spots to increase their body temperatures (Moore 2002).

Sunning may kill lice by desiccating them directly, or by increasing the vulnerability of lice to preening as they move around on the body to escape the heat. Moyer and Wagenbach (1995) exposed lice on model black noddy (*Anous minutus*) wings to sunny and shady microhabitats. The duration of exposure was typical of natural sunning bouts, and the temperature of the models was similar to that of sunning live noddies. Significantly more lice died in the sun than shade, showing that exposure to sun can kill lice, even when preening is not involved. Blem and Blem (1993) compared the rate of sunning by fumigated and non-fumigated violet-green swallows (*Tachycineta thalassina*). Fumigated swallows sunned less than controls, suggesting that the motivation to engage in sunning decreases when ectoparasites are not present.

These experiments suggest that sunning helps to combat lice. However, additional work is needed to test the number of lice killed by sunning under natural conditions. It would be interesting to know whether sunning is more effective in controlling lice on birds with dark plumage, compared to birds with light plumage. Preliminary work suggests that dark feathers heat up more rapidly than white feathers—and to higher temperatures—and that lice move more quickly on dark feathers than light feathers at high temperatures (Clayton et al. 2010). It is tempting to speculate that one cost associated with the evolution of light plumage is a reduction in the effectiveness of sunning for the control of lice.

Interspecific cleaning

Medium- to large-bodied mammals are groomed by 100 species of birds belonging to 32 families (Sazima 2011). These "cleaner" birds feed on ticks, flies, and the skin and blood of their "clients." There are also records of cleaner birds consuming small numbers of lice (Moreau 1933). However, it seems unlikely that cleaner birds play a major role in combating lice or other small ectoparasites on most mammals. The most interesting report of interspecific grooming of lice is Nuttall's (1917) observation that baboons were "hired out" to control head lice on people in Lisbon in the eighteenth century. This account suggests an amusing solution to the current epidemic of head lice in school kids (chapter 2).

So far in this chapter we have reviewed mechanisms for resisting lice that include chemical defense, molt, dusting, sunning, anointing, nest fumigation, and several forms of grooming. The mechanisms evolved by a particular group of hosts should vary, depending on the composition of their ectoparasite community. Birds with head lice should experience louse-mediated selection for efficient scratching. However, birds without head lice, such as pigeons and doves, should not. It would be interesting to compare the diversity of anti-parasite adaptations among orders of birds or mammals to the diversity of their lice and other ectoparasites. Some anti-parasite behaviors may interact—for example, sunning appears to increase the effectiveness of preening (Moyer and Wagenbach 1995). Some behavioral mechanisms are also known to interact with host immune responses, which we consider next.

Immunological resistance
Mammals

Studies of domesticated and laboratory mammals demonstrate that immunological responses can provide effective resistance against lice and other ectoparasites (Nelson et al. 1977; Nelson 1987; Bany et al. 1995; Baron and Weintraub 1987; Wikel 1982, 1984, 1996, 1999; Lehane 2005; Owen et al. 2010). However, immune responses are complex and influenced by many factors, such as host age, nutrition, and stress (Nelson et al. 1977; Marshall 1981a; Nelson 1984; Tschirren et al. 2006; Rueesch et al. 2012). These complexities aside, immune responses combat lice and other ectoparasites in one of two main ways. The first is for acquired resistance in response to parasite antigens to lead to inflammation, epidermal thickening, vasoconstriction, and/or a decrease in blood vessel number. These changes impair the feeding ability of lice, causing their populations to crash (fig. 4.11). The second mechanism is for acquired resistance to trigger directed grooming in response to hypersensitivity and itching (Lehane 2005). The two mechanisms

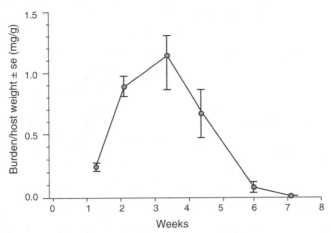

FIGURE 4.11. Effect of acquired immunity on the mean (± se) density of sucking lice (*Polyplax serrata*) on eight lab mice with impaired grooming. Mice (Cox/Swiss) were initially infested with several hundred lice each. After a rapid increase, the louse populations crashed due to systemic immune responses that interfered with their feeding ability. Louse density was measured using the ratio of louse mass (mg) to host mass (g); 1 mg of lice equaled 15–20 fresh, active lice. After Ratzlaff and Wikel (1990).

can interact, as when inflammation at the site of infestation forces lice to abandon regions normally protected from grooming, such as the back of the neck, and move onto regions that are more vulnerable to grooming (Lehane 2005).

The effectiveness of acquired resistance, even in the absence of grooming, was demonstrated long ago using lab mice experimentally infested with sucking lice (*Polyplax serrata*) (Nelson et al. 1977; Lehane 2005). Bell et al. (1966) showed that localized resistance is associated with dramatic reductions in lice. Bell et al. (1982) later demonstrated that resistance can be transferred between mice by skin grafting. More recently, Ratzlaff and Wikel (1990) showed that some host responses are systemic—that is, they can be elicited in body regions well removed from the site of louse feeding. The authors also showed that systemic reactions can be induced by "immunizing" mice with soluble components of crushed lice. Immunized mice reduced their louse loads by 62%, although this was less than the 94% reduction in mice with naturally acquired immunity (Ratzlaff and Wikel 1990). Systemic reactions in mice are similar to those in people who experience rashes on their chests, backs, and abdomens after they allow body lice to feed on their forearms (Hirschfelder and Moore 1919).

Ratzlaff and Wikel (1990) documented proliferative cell responses in the lymph nodes (but not spleens) of mice within eight days of experimental infestation. These results, together with the inability of athymic mice (lymph nodes removed) to develop resistance, suggest that T cells in draining lymph nodes mediate acquired resistance to lice. When the proliferation of T lymphocytes was suppressed early in the infestation, reductions in lice were delayed, further suggesting that a cell-mediated mechanism is required for the expression of resistance (Nelson et al. 1983). Several host cell types, including neutrophils, eosinophils, and lymphocytes, are known to proliferate at sites of attachment by lice (Nelson et al. 1977, 1979).

Experimental studies of other laboratory and domesticated mammals also reveal immune responses to lice. Examples include work on rats (Volf 1991, 1994), sheep (James and Moon 1998; James 1999), cattle (Nelson et al. 1970; Colwell and Himsl-Rayner 2002), and rabbits infested with rabbit-adapted strains of human body lice (*P. h. humanus*) (Mumcuoglu et al. 1997; Ben-Yakir et al. 2008). Some of these studies show that immune responses can occur even when lice do not feed on blood. For example, there is evidence of hypersensitive immune responses to the sheep chewing louse *Bovicola ovis*, which ingests lipid, scurf (dandruff), dead skin flakes, and bacteria, but not blood. Effective immune responses by experimentally infested sheep include both cellular responses and serum antibodies, and they may also involve secretion of immunoglobulins on the surface of the host's skin.

To our knowledge, immune responses of wild mammals to sucking or chewing lice have not been measured.

Birds

Birds are known to combat ectoparasites, such as blood-feeding mites, with acquired immune responses (Owen et al. 2010). These responses can damage parasite tissues through the release of proteolytic compounds by granulocytes (Wikel 1996; Owen et al. 2009). Host tissue swelling (edema) helps defend hosts by reducing the size of parasite blood meals. Immune responses decrease parasite fecundity, inhibit parasite molting, and increase parasite mortality (Dusbábek and Skarkova-Spakova 1988; Owen et al. 2009). Most work on the effectiveness of immune responses in combating ectoparasites of birds has been done with poultry ticks, mites, bugs, and parasitic flies (Owen et al. 2010). Little or no work has been done with bird lice.

In contrast, some work has been done on the role of immune responses in combating the lice of wild birds. However, most of this work involves testing for correlations between "immunocompetence" and louse load (Saino et al. 1995; Eens et al. 2000; Blanco et al. 2001; Fairn et al. 2012). Many of these studies were designed to estimate the overall strength of the immune system using general adjuvants such as phytohaemagglutinin (PHA), rather than actual parasite antigens. An advantage of these studies is that they control for differences in the unknowable exposure histories of wild birds. A disadvantage of such studies is that they do not provide information on particular immune responses to specific parasites (Owen and Clayton 2007; Huber et al. 2010).

Møller and Rózsa (2005) reported a positive correlation between the immunocompetence of nestlings of different species of altricial birds and the taxonomic richness of their amblyceran lice. Interestingly, they found no correlation between immunocompetence and the richness of ischnoceran lice. The authors suggested that this difference is due to the fact that amblyceran lice encounter the immune defenses of birds because they feed on blood, while ischnoceran lice do not. They further suggested that the positive correlation of Amblycera and nestling immunocompetence reflects greater niche specialization by lice on birds with strong immune responses. They cautioned, however, that the causal arrow may be reversed—that is, greater amblyceran richness could lead to stronger immune responses.

Whiteman et al. (2006) reported a negative correlation between natural antibody (NAb) levels and the abundance of amblyceran lice (*Colpocephalum turbinatum*) on Galápagos hawks (*Buteo galapagoensis*). As in Møller and Rózsa's (2005) study, there was no correlation between immunocompetence and ischnoceran lice (*Deegeriella regalis*). This study provided additional evidence indicating the two suborders of lice do, in fact, differ in their vulnerability to the host's immune system.

It is important to bear in mind that the diets of most ischnoceran lice have not been studied carefully (Fairn et al. 2012). Some species of Ischnocera may feed opportunistically on living skin or blood (Marshall 1981a; Lehane 2005). Moreover, activation of the immune system may not require contact with living skin or blood, as in the case of sheep lice mentioned above (James 1999; Volf 1994). Therefore, the hypothesis that Amblycera are more vulnerable than Ischnocera to host immune responses requires further study.

Genetics of resistance

Heritable resistance to pathogens and other parasites is common, and more genetically diverse host populations tend to be more resistant (Wakelin and Apanius 1997; Schmid-Hempel 2011). Heritable resistance to ectoparasites is no exception, at least in the limited number of cases where it has been studied (Nelson et al. 1977; Lehane 2005; Luong et al. 2007; Owen et al. 2010). Hosts show considerable variation within populations in their susceptibility and response to ectoparasitism. This variation ranges from individuals that fail to respond to parasite bites, to individuals that develop degrees

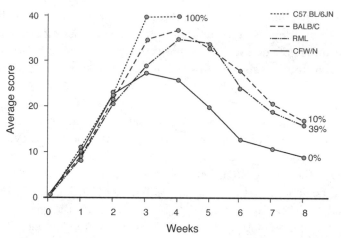

FIGURE 4.12. Comparative effects of sucking lice (*Polyplax serrata*) on experimentally infested strains of lab mice (28–30 mice per strain). Grooming was blocked; thus, host defense was largely immunological. Lice increased on all strains over the first month of the experiment. Subsequent effects of lice on the different strains varied dramatically, with percentages showing host mortality. All individuals in one strain (C57 BL/6JN) died within a month, while none of the mice in another strain (CFW/N) died. Louse populations decreased on all three of the surviving strains of mice over the second month of the experiment. Lice were scored as follows: 0 = no lice, 10 = rare, 20 = few to moderate, 30 = many, and 40 = very many. After Clifford et al. (1967).

of swelling and itching, to individuals that die outright from anaphylactic shock in response to a single bite (Lehane 2005). Variation of this kind has been shown to correlate with underlying genetic variation in laboratory mammals.

Clifford et al. (1967) demonstrated variable responses among genetic strains of laboratory mice experimentally infested with *P. serrata* sucking lice. The responses ranged from 100% mortality of mice within four weeks, to no mortality and the near elimination of lice within eight weeks (fig. 4.12). Mortality of mice was apparently due to severe anemia (Stewart et al. 1976). Subsequent breeding trials indicated that resistance had a heritable basis. Ratzlaff and Wikel (1990) and Volf (1991) also showed genetic variation in the immune responsiveness of mice to sucking lice (*P. serrata* and *P. spinulosa*).

Heritable resistance of hosts to lice has also been reported in domesticated mammals, such as sheep (James et al. 2002). Pfeffer et al. (2007) compared the heritability of resistance to *Bovicola ovis* chewing lice among four breeding lines of sheep that differed in their susceptibility. The authors concluded that selective breeding for increased resistance might be an effective control method.

Aspects of grooming behavior are also known to be under genetic control. Greer and Capecchi (2002) showed that mice with experimental disruption of the *Hoxb8* gene dramatically increased rates of self-grooming and mutual grooming of littermates. However, the effects of such changes on ectoparasite loads have not been tested. Strong effects of grooming on lice of laboratory mice (Lodmell et al. 1970) make it likely that increases in grooming by *Hoxb8* knockouts would lead to decreases in parasite load. Resistance to ectoparasites by other mammals, such as cattle, has been attributed to better grooming "arousal" behavior (Nelson et al. 1977).

In contrast to mammal lice, work on the genetics of resistance to bird lice is preliminary. Møller et al. (2004b) demonstrated an additive genetic component underlying possible similarities in the louse loads of adult barn swallows and their offspring. To do this they used cases of extra-pair paternity to partition environmental and genetic effects on parent-offspring correlations in feather hole number (but see box 3.1).

Several studies have tested the hypothesis that genetically uniform bird populations are more susceptible to lice than genetically diverse populations. Whiteman et al. (2006) showed that inbred populations of Galápagos hawks on smaller islands have more amblyceran lice. These populations also had lower natural antibody (NAb) levels, suggesting that inbreeding reduces resistance. In contrast, Hoeck and Keller (2012) found no relationship between inbreeding, louse abundance, and different measures of immunocompetence among 14 populations of Galápagos mockingbirds (*Mimus* spp.). As the authors pointed out, however, the repeatability of their measures of immunocompetence, and the statistical power of the overall study, were limited.

In a study of lesser kestrels (*Falco naumanni*), Ortego et al. (2007) demonstrated a strong correlation between host homozygosity and the prevalence of the single species of louse present (fig. 4.13). The effect was independent

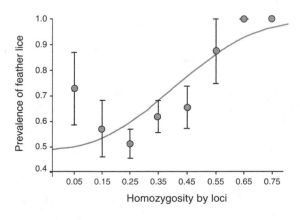

FIGURE 4.13. Mean (± se) prevalence of lice (*Degeeriella rufa*) in relation to the homozygosity of 181 lesser kestrels in central Spain. Genetic diversity was assessed at 11 microsatellite loci. The solid line shows predicted values from a statistical model. After Ortego et al. (2007).

of other factors correlating with prevalence, including host sex, location, year, and colony size. Because this result was a genome-wide effect, with no single locus contributing disproportionately, the authors concluded that overall genetic variation was responsible. The only species of louse on the kestrels was a feather-feeding ischnoceran, which presumably did not interact with the immune system. The authors concluded that more susceptible birds may have had less stamina, and were therefore less capable of investing energy in behavioral defenses, such as preening. Genetic variation has been shown to correlate with behavioral defenses against ectoparasites in other systems (Luong et al. 2007).

In summary, resistance mechanisms of mammals and birds have a heritable basis and are presumably capable of evolving in response to selection imposed by lice and other ectoparasites. These data are consistent with the hypothesis that parasite-mediated mate choice may combat lice (chapter 3). In addition to avoiding direct infestation, the choice of a louse-free mate may increase the probability of transmitting "good genes" for parasite resistance to offspring.

5   COUNTER-ADAPTATIONS OF LICE

We see these beautiful co-adaptations ... in the humblest parasite
which clings to the hairs of a quadruped or feathers of a bird.
—Darwin 1859

Lice and other permanent parasites complete all stages of
their life cycle on the body of the host. This "host or bust"
strategy provides ready access to food and shelter, but it
comes at a cost. Over evolutionary time, lice have lost the
ability to disperse easily away from the host, or to survive
for long off the host. They have evolved morphological, physiological, and
behavioral adaptations for 1) attachment to the host, 2) locomotion on the
host, and 3) escape from host defense. The three classes of adaptations are
intertwined; however, the third category includes the most apparent and
best-studied counter-adaptations. We begin by considering adaptations
for attachment and locomotion. We then consider a variety of counter-
adaptations coevolved by lice for circumventing host defense.

Attachment and locomotion

Lice and other ectoparasites have morphological adaptations that facilitate
attachment to, and locomotion on, the host. These adaptations include
hooks, clamps, suckers, and overall body size and shape (Gorb 2001;
Burkett-Cadena 2009). For example, the laterally compressed shape of fleas
facilitates movement through "forests" of fur or feathers. Similarly, the dorsoventrally compressed shape of lice appears to serve the same function.
Many fleas have spines and combs (ctenidia) that may facilitate attachment
to the host, especially during grooming sessions (fig. 5.1a; Humphries 1967;
Krasnov 2008). The most compelling evidence in the support of this hypothesis is the tight correlation between host hair diameter and spacing of
the ctenidial (comb) "teeth" in some fleas (fig. 5.1b; Humphries 1967; Amin
and Wagner 1983).

Like fleas, lice have spine-like setae and combs that may improve tenac-
ity on the host. Some groups of lice have comb-like spines that may contribute to attachment (fig. 5.1c). Other groups are densely covered with setae
(fig. 5.1d, plate 2c). Some lice have intricate combs on their legs (fig. 5.1e).
Kéler (1952) hypothesized that tarsal combs facilitate movement of lice
among host feathers. He speculated that the combs increase traction when
lice walk or run through the plumage. Spine-like setae and combs may also

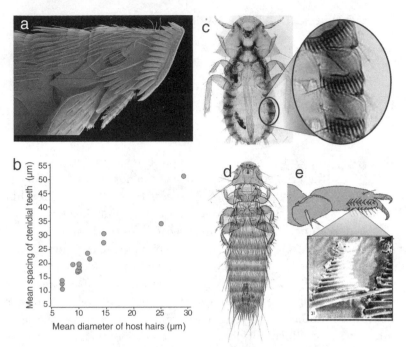

FIGURE 5.1. Adaptations related to attachment and locomotion. (a) Head of a flea (*Stephanocircus dasyuri*) that parasitizes bandicoots (*Thylacis obesulus*). The spines and combs (ctenidia) are thought to facilitate attachment to the host. (SEM by M. W. Hastriter.) (b) Relationship between the spacing of ctenidial teeth and host hair diameter across fleas and their mammalian hosts (after Humphries 1967). (c) Abdominal combs on the chewing louse *Rhopaloceras rudimentarius*, from a little tinamou (*Crypturellus soui*) (photo by V. S. Smith, inset photo by D. R. Gustafsson). (d) Dense long hairs (setae) on a chewing louse (*Amyrsidea ventralis*) from an Argus pheasant (*Arqusianus argus*) (photo by V. S. Smith). (e) Rows of comb-like setae on the tarsus of *Colpocephalum* sp., a genus of chewing louse that occurs on many species of birds (SEM from Clay 1969).

improve the ability of lice to avoid removal by grooming hosts (Johnson and Clayton 2003a). Because setae can also have a sensory function, they may help lice monitor and dodge grooming (Rothschild and Clay 1957; Smith 2001; Cruz and Mateo 2009). Generally speaking, however, the functional significance of combs, setae, and spines has not been tested. Microsurgical ablation or other experimental manipulations could be informative in testing the role of these structures in attachment, locomotion, and escape.

On a broader comparative scale, the overall body size of most lice is correlated with that of their hosts (Johnson et al. 2005; Cannon 2010). This relationship, called "Harrison's rule" in honor of the first investigator to describe it (Harrison 1915), holds for a diversity of groups, including para-

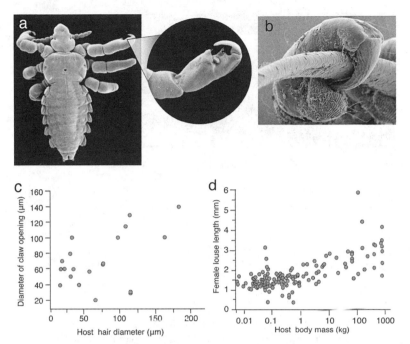

FIGURE 5.2. Harrison's rule in sucking lice. (a) *Haematopinus bufali* from an African buffalo (*Syncerus caffer*) with inset of tarsal claw (from Turner et al. 2004); (b) Close-up of a human head louse *(Pediculus humanus capitis)* gripping a human hair (SEM by M. Turner); (c) Relationship of the grasping diameters of lice claws to host hair diameters (n = 20 species; r = 0.59, P = 0.01); (d) Relationship of louse size to host size across ten orders of mammals (Macroscelidea, Tubulidentata, Scandentia, Primates, Lagomorpha, Soricomorpha, Carnivora, Perissodactyla, Artiodactyla, and Rodentia) (n = 147 species, r = 0.53, P < 0.001) (c and d after Cannon 2010).

sitic worms, crustaceans, fleas, flies, lice, and ticks, as well as herbivorous aphids, thrips, beetles, flies, moths, and flower mites (Harvey and Keymer 1991; Kirk 1991; Thompson 1994; Poulin 2007; Sasal et al. 1999; Morand et al. 2000; Johnson et al. 2005). The adaptive significance of matching size varies, depending on the host-parasite relationship in question.

Sucking lice show evidence of Harrison's rule (fig. 5.2). The phenotypic interface mediating this relationship is the match between host hair diameter and the grasping diameter of the parasite's claw (fig. 5.2a-c). Host hair diameter is correlated with overall size of the host (Reed et al. 2000a), and claw diameter is presumably correlated with the overall size of the louse, although the latter has not been tested to our knowledge. The end result is that overall louse size and host size are correlated (fig. 5.2d). The match between claw size and hair size is undoubtedly important for

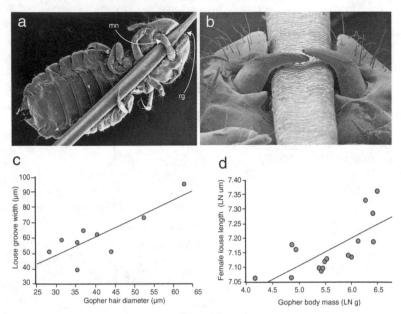

FIGURE 5.3. Harrison's rule in chewing lice of mammals. In this case the lice attach by slipping a hair through the rostral groove and gripping it with their mandibles. (a) *Damalinia ornata* from an African hartebeest (*Alcelaphus buselaphus*) showing the rostral groove (rg) and mandibles (mn) gripping a hair (SEM by M. Turner); (b) Close-up of *Damalinia crenelata* gripping the hair of an African blesbuk (*Damaliscus dorcas*) with the mandibles (from Turner 2003); (c) Rostral groove width relative to pocket gopher hair diameter; regression on independent contrasts was positive ($R^2 = 0.501$, $P = 0.033$); (d) Relationship of louse body size to pocket gopher size; regression on independent contrasts was positive ($R^2 = 0.808$, $P = 0.0004$) (c and d after Morand et al. 2000).

attachment and locomotion. This is apparent, for example, in the case of human head lice and pubic lice, which have claw sizes that are correlated with the diameters of head and pubic hair, respectively (fig. 2.8). The match between hair and claw sizes presumably also helps lice avoid removal by host grooming.

Mammalian chewing lice also show evidence for Harrison's rule (fig. 5.3), but they attach to the host differently. Pocket gopher lice use robust chewing mandibles to grasp individual hairs, and they have a rostral groove through which the hair passes (fig. 5.3a,b). The width of this groove in pocket gopher lice is correlated with host hair diameter (fig. 5.3c), suggesting that it plays an important role in attachment (Reed et al. 2000a). Overall louse size and host size are also correlated (fig. 5.2d).

Reed and Hafner (1997) transferred host-specific lice between species of captive pocket gophers. The ability of lice to establish on novel host species was correlated with host phylogenetic relatedness, which was correlated with host size (Reed et al. 2000a). A mismatch between rostral groove size and gopher hair diameter may have contributed to the inability of lice to establish on pocket gopher species of the "wrong" size. It would be interesting to know just how much the correct match in groove size contributes to the ability of lice to remain attached to their normal hosts.

Although matching components of size are clearly adaptive for lice, they are probably not coadapted traits, per se, because mammals are unlikely to have evolved different hair diameters to resist lice or other ectoparasites. Hair is presumably under selective constraints related to thermoregulation and other functions. This is not to say that other forms of host defense are not coadapted with the morphology and behavior of lice. Indeed, several lines of evidence indicate that morphological and behavioral components of grooming have coadapted with lice (chapter 4). Coadaptation is most apparent between birds and their feather lice, which we consider next.

Counter-adaptations involving body size

> The size of an organism is under constant selective surveillance.....
> Size rules life.
> —Bonner 2006

Johnson et al. (2005) tested for Harrison's rule across 78 species of bird lice, representing several dozen genera (from Amblycera and Ischnocera). The authors showed that, despite the fact that avian plumage is more diverse in structure than mammal hair, bird lice also show evidence of Harrison's rule (fig. 5.4). The effect of plumage diversity on the evolution of bird lice is illustrated by differences in the morphology and behavior of genera specialized for different microhabitats on the body of the host. For example, the large flight feathers of the wings and tail have ridges on their surfaces that are formed by feather barbs, which connect to adjacent barbs with zipper-like barbules. The interbarb spaces create furrows into which cigar-shaped wing lice can insert themselves to avoid preening (fig. 5.5a-c). In contrast, more oval-shaped body lice live in the abdominal contour feathers of the host, where they avoid preening by dropping between adjacent feathers, or by burrowing through the downy regions of the feathers (fig. 5.5d). Size influences the ability of both body lice and wing lice to escape from simulated host preening (Bush and Clayton 2006).

Escape from host defense by ecological replicate lice

Matching size at the phenotypic interface, such as the fit between wing lice and the interbarb space (fig. 5.5b), is consistent with coadaptation

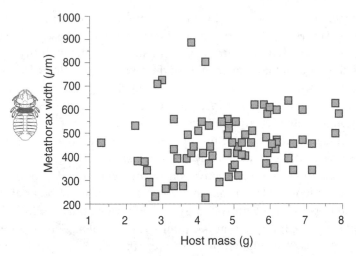

FIGURE 5.4. Harrison's rule in chewing lice of birds. The plot includes a variety of amblyceran and ischnoceran lice genera from 23 (10%) of the world's bird families (from five orders). Louse size (metathorax width, shown in grey) and bird mass are significantly correlated; a regression through the origin of phylogenetically independent contrasts revealed a significantly positive association (P = 0.018). After Johnson et al. (2005).

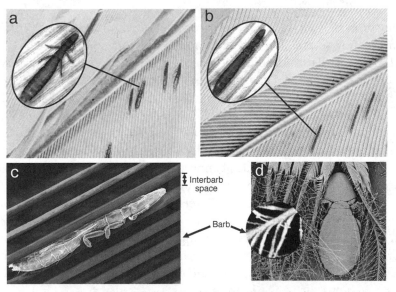

FIGURE 5.5. Behavior of rock pigeon wing lice (*Columbicola columbae*) and body lice (*Campanulotes compar*): (a) Wing lice on the undersurface of a flight feather, and (b) inserted into the interbarb spaces in response to simulated preening; (c) Modified SEM of insertion behavior; (d) Body louse in the downy matrix of abdominal contour feathers (a,b by S.E.B.; c by A. Bartlow and S. Villa; d after Johnson and Clayton 2003b).

by the lice in response to preening-imposed selection. It is conceivable, however, that matching size is due to ecological tracking (see chapter 1), in which lice establish populations on birds of the "correct" size, rather than coevolving with them. For this reason, Clayton et al. (1999) performed an experiment to test whether preening exerts selection on feather lice. Lousy rock pigeons were fitted with bits (fig. 4.4) for several months to relax any preening-imposed selection. Over this period, lice on the birds were expected to increase in size because large individual insects tend to be more fecund, all else being equal (Sibly and Calow 1986). After taking samples of the lice (Time 1), bits were removed from experimental birds, but left in place on control birds. The lice were then sampled again after several weeks (Time 2) and their sizes compared (fig. 5.6). Time 2 lice from preening birds were significantly smaller than Time 1 lice, whereas lice on bitted birds did not change significantly in size over time. The results of this experiment showed that preening did, in fact, select for small body size in the lice. Small size presumably increased the ability of lice to escape from

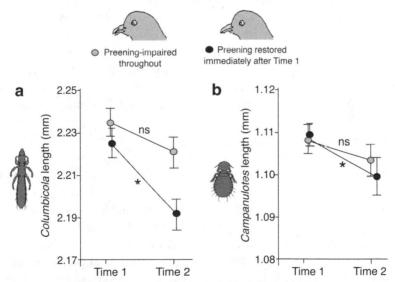

FIGURE 5.6. Selective effect of preening on rock pigeon (a) wing lice (*Columbicola columbae*) and (b) body lice (*Campanulotes compar*). Birds were bitted for several months prior to the experiment. Samples of lice were collected from birds at Time 1, then the bits of experimental birds were removed to restore their preening ability. Bits on control birds were left in place. Lice were re-sampled from all birds about 26 days later (Time 2). Body length (mean ± se) decreased significantly in populations of wing lice (P < 0.002) and body lice (P < 0.05) on experimental birds, but not on control birds (ns = no significant difference). After Clayton et al. (1999).

preening by increasing the efficiency of insertion in the case of wing lice, and increasing the efficiency of burrowing in the case of body lice.

Bush and Clayton (2006) conducted a more direct test of the influence of body size on the ability of lice to escape from host preening. They transferred lice among pigeons and doves of different sizes and quantified the population sizes of the lice after a period of two months (about two louse generations). Twenty-five wing lice and 25 body lice from rock pigeons were transferred to each of twelve individuals of each of four novel host species (fig. 5.7a,b). Twenty-five wing lice and 25 body lice from common ground doves (*Columbina passerina*) were also transferred to each of twelve individuals of each of four novel host species (fig. 5.7c,d). Half of the birds in each species were bitted to impair preening ability, while the other half had normal preening ability. Three of the four novel hosts getting rock pigeon lice (fig. 5.7a,b) were smaller in body size than the rock pigeons. Three of the four novel hosts getting common ground dove lice (fig. 5.7c,d) were larger in size than the common ground doves.

On birds that could preen normally, transferred lice could establish viable populations only when host species were similar in size to the native host (fig. 5.7). Most populations of lice transferred to novel (preening) hosts went locally extinct, or nearly so, by the end of the experiment.

In contrast, on birds with impaired preening, lice were able to establish viable populations on smaller-bodied hosts: wing and body lice populations from rock pigeons increased on all species of hosts (fig. 5.7a,b). This experiment confirmed that preening interacts with small host body size to exert strong selection on lice. The interaction between preening and body size effectively reinforced the host specificity of both wing and body lice. Note that the lice were fully capable of finding food and other resources needed to survive and reproduce on a range of (preening-impaired) host sizes over the two generations.

Lice transferred to *larger* novel hosts were unable to establish viable populations, *even when preening was impaired* (fig. 5.7c,d). Related experiments show that the lice are capable of feeding on, and remaining attached to, feathers from these large hosts. The only measure of performance that decreased on large feathers was running speed (Bush 2004). However, more work is needed to test whether the effect on running speed was responsible for the inability of lice to establish viable population on large novel hosts, even when those hosts were bitted.

Bush et al. (2006b) further explored the insertion ability of wing lice in relation to host size. Lice placed on (detached) feathers from hosts much smaller than the native host could not insert fully (fig. 5.8). Although lice still showed insertion behavior, their abdomens protruded, making them

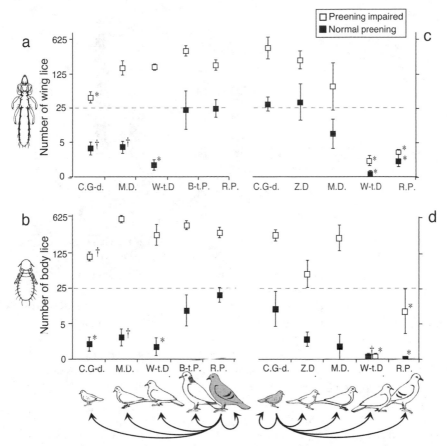

FIGURE 5.7. Population sizes (mean ± se) of lice transferred to novel host species (note log scale on y-axis): (a) *Columbicola columbae*; (b) *Campanulotes compar*; (c) *Columbicola passerinae*; and (d) *Physconelloides eurysema*. In the first experiment, wing lice (a) and body lice (b) from rock pigeons were transferred to progressively smaller host species, and to rock pigeon controls (in grey). In the second experiment, wing lice (c) and body lice (d) from common ground doves were transferred to progressively larger host species, and to common ground dove controls (in grey). Dashed lines show the number of lice transferred to each bird at the start of the experiment, which was two months in duration (about two lice generations). Populations that differed significantly from controls (within preening treatments) are indicated by *P ≤ 0.01, and †P ≤ 0.05. Host species (drawn to scale) varied in body size (grams) by nearly an order of magnitude: C.G-d., common ground dove (*Columbina passerina*, 45 g); Z.D., zebra dove (*Geopelia striata*, 50 g); M.D., mourning dove (*Zenaida macroura*, 113 g); W-t.D., white-tipped dove (*Leptotila verreauxi*, 177 g); B-t.P., band-tailed pigeon (*Patagioenas fasciata*, 353 g); R.P., rock pigeon (*Columba livia*, 364 g). After Bush and Clayton (2006).

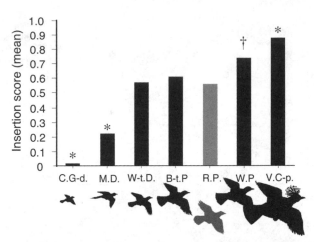

FIGURE 5.8. Insertion ability of rock pigeon wing lice (*Columbicola columbae*) is related to host size. Bush et al. (2006b) placed lice on (detached) #5 primary wing feathers of six novel host species (drawn to scale in black), and native host controls (grey). The position of the lice was scored on a scale from 0, for lice on the surface of a feather, to 1.0 if the lice were fully inserted (fig. 5.5b). Lice inserted significantly less on feathers of small dove species, compared to controls, and they inserted significantly more on feathers of the largest pigeon species (Dunnett's post hoc *P < 0.01, †P < 0.05). Host names as in fig. 5.7, plus W.P. = wood pigeon (*Columba palumbus*), and V.C-p. = victoria crowned-pigeon (*Goura victoria*). The hosts ranged in size from 30 g (C.G-d.) to 2,400 g (V.C-p), with rock pigeon controls weighing about 350 g (Dunning 1993; del Hoyo et al. 1997). After Bush et al. (2006b).

vulnerable to preening. Incomplete insertion was also noted in lice transferred to smaller hosts in the experiments described above with live birds (fig. 5.7). Interestingly, lice placed on (detached) feathers from large novel hosts could insert even more completely than lice on their native hosts. Why, then, do different species of *Columbicola* tend to match the size of their hosts (fig. 5.9a), rather than remaining small? The answer may lie in the correlation between insect body size and fecundity mentioned above. Although small size may be advantageous for avoiding preening, it does not necessarily maximize fitness in other respects. In summary, insertion by wing lice is a coadaptation to avoid host preening. Experiments testing other possible adaptive functions of insertion have produced negative results (box 5.1).

Given the critical role of size in preening avoidance, one might predict that both wing lice and body lice should follow Harrison's rule. This is not the case, however; wing lice show evidence for Harrison's rule (fig. 5.9a), but body lice do not (fig. 5.9b). Harrison's rule in wing lice makes sense because wing louse size is correlated with interbarb size (fig. 5.9c), which is, in turn,

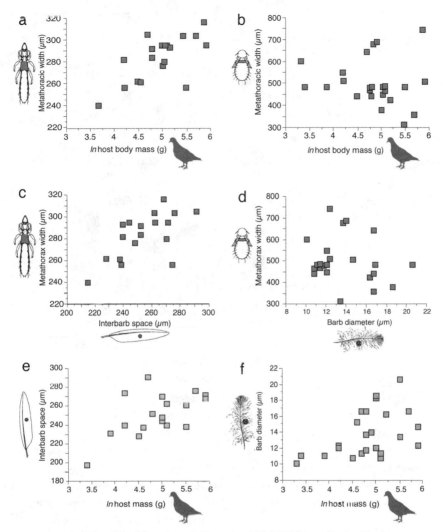

FIGURE 5.9. Relationship of female metathoracic width (FMW) (grey) to host body mass for (a) 19 lineages of columbiform wing lice, and (b) 24 lineages of columbiform body lice (25 species of hosts in each case). A regression through the origin of phylogenetically independent contrasts for wing lice revealed a strong positive association (P = 0.001). There was no significant association for body lice (P = 0.30); (c) FMW for wing lice relative to interbarb space (P = 0.001); (d) FMW for body lice relative to barb diameter (P = 0.17); (e) interbarb space relative to host body mass (P = 0.006); (f) barb diameter relative to host body mass (P = 0.70). After Johnson et al. (2005).

BOX 5.1. Insertion is not required for attachment by wing lice

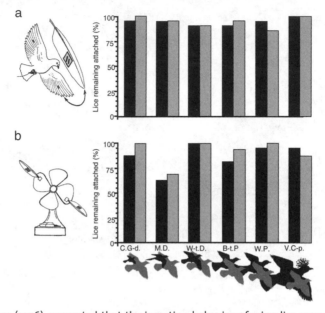

Stenram (1956) suggested that the insertion behavior of wing lice may also be an adaptation for remaining attached to the feathers of flying birds. To test this hypothesis, Bush et al. (2006b) tested whether rock pigeon wing lice can remain attached to the feathers of different-sized hosts, including feathers that are too small to allow insertion (fig. 5.8). Squares (1 cm²) of feather vane were cut from the #5 primary wing feather of each of several species of pigeons and doves. The squares were then grafted onto the #5 primary feathers of live rock pigeons and held in place with (inert) three-dimensional paint that was used to form a "corral" around each feather graft. Each pigeon had a feather with a novel species graft attached to one wing, and a feather with a control graft from another rock pigeon attached to the other wing. Birds were tethered on a long line and allowed to fly a distance of 50–100 meters after placement of lice on the feather grafts. The grafts were checked to determine whether the lice were still attached (a). Surprisingly, nearly all lice remained attached to feather grafts, regardless of size (see final paragraph for details).

To test the relationship between feather size and attachment ability under more strenuous conditions, experimental and control feathers were removed from pigeons and taped to the blades of a high-speed fan. Lice were placed on the grafts and the fan turned on for 20 minutes at speeds simulating pigeons in level flight (Johnston and Janiga 1995). Although fewer overall lice remained attached to the feathers (b), there was still no significant difference in the number of lice on experimental versus control grafts. Thus, insertion was not required for wing lice to remain attached to feathers. The reason is that wing lice are capable of remaining attached to feathers by locking onto them with their mouthparts. When barbules are blocked with fingernail polish, the lice can no longer remain attached (Bush et al. 2006b).

Rock pigeon wing lice (*C. columbae*) were capable of remaining attached to feather grafts from hosts varying in size from common ground doves (45 grams) to Victoria crowned pigeons (2,400 g). This was the case even when the lice could not insert themselves between the interbarb spaces of feathers (fig. 5.8, mourning doves and common ground doves). Bars show the percent of wing lice remaining attached to feather grafts from six novel host species (black) and the native host (gray). Attachment did not differ significantly among host species on (a) feather grafts on flying rock pigeons, or (b) feather grafts on the blades of a high-speed fan. See fig. 5.7 for host names (after Clayton et al. 2003b; Bush et al. 2006b).

correlated with overall host size (fig. 5.9e). It is perhaps not surprising that body lice do not follow Harrison's rule, given that body lice are not correlated with body feather size (fig. 5.9d), nor is body feather size correlated with host size (fig. 5.9f).

What is not yet clear is why size still plays a role in the ability of body lice to escape from host preening (fig. 5.7 b). The answer may be related to the influence of some unmeasured component of feather size on running speed, burrowing speed, or another component of body louse performance. In summary, the relationship between louse body size, feather size, and overall host body size is complex.

Counter-adaptations involving color

> He told me about the incredible artistic wit of mimetic disguise.
> —Nabokov 1963

Cryptic coloration of predators to avoid detection by prey—and of prey to avoid detection by predators—has long been one of the most compelling examples of evolution by natural selection (Darwin 1859; Cott 1940; Ruxton et al. 2004; Stevens and Merilaita 2009). Recent work shows that cryptic coloration also evolves in external parasites. For example, geometrid moth larvae (*Ematurga atomaria*) must avoid predation when they feed on host plants, yet the larvae feed on species of plants that differ in color. Moths are capable of developmental changes in coloration via phenotypic plasticity. Larvae reared on dark-colored host plants are dark, while larvae reared on light-colored host plants are light (Sandre et al. 2013).

In the case of permanent parasites, such as lice, the host represents both the background the parasite must match, as well as an agent of selection for crypsis to escape from host preening. Cryptic coloration was first noticed in human lice. Darwin (1871) commented on variation in the color of head lice among human races, suggesting that the color of lice may match hair color. Murray (1861) published similar observations a decade before Darwin. In another example, a biologist working in South America noted "from personal experience" that crab lice transferred from a black person to a white person get lighter in coloration over the course of several louse generations (Nuttall 1919a).

Fahrenholz (1915) classified human body lice into different subspecies partially on the basis of color. Nuttall (1919b) questioned this practice and showed that the color of body lice depends on the color of the background on which they are reared. Later experiments by Busvine (1946) demonstrated that coloration in body lice has a heritable component (fig. 5.10). More recently, Veracx et al. (2012) documented extensive variation in the coloration of human body lice from around the world. Lice in the study were photographed to quantify color, then genotyped using four intergenic spacers (S2, S5, PM1, PM2). Phylogenetic reconstruction showed that color is not a conserved trait, but is evolutionarily labile across the tree (Veracx et al. 2012). It would be interesting to use the human body louse genome (Pittendrigh et al. 2009) to explore the genetic architecture of color in human lice.

Crypsis as a counter-adaptation

Bird lice also vary in color and show strong evidence for the evolution of crypsis. Dark-colored bird species tend to have dark-colored lice, whereas light-colored bird species tend to have light-colored lice (Rothschild and Clay 1957). Bush et al. (2010) used phylogenetically independent comparisons to test for the evolution of crypsis by lice over macroevolutionary time. The authors compared the color of congeneric lice found on closely related birds that were mostly white or black, such as light and dark species of swans, or light and dark species of cockatoos (plate 12). The results showed a significant correlation between host and parasite color (fig. 5.11a). In contrast, head lice, which are not affected by preening, showed no evidence of crypsis (fig. 5.11b). Thus, head lice are the exception that proves the rule. Most head lice are darkly colored, indicating that coloration is not a plastic response to feeding on different colored feathers. It would be interesting to test whether cryptic coloration is most pronounced in lice

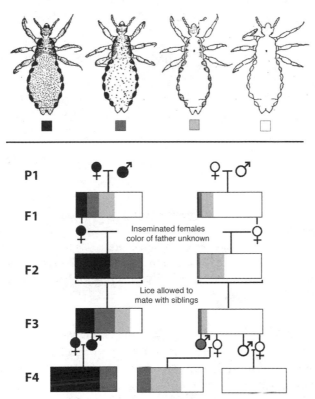

FIGURE 5.10. Heritability of color in human body lice (*Pediculus humanus*). Lab-reared lice were separated into four categories based on color (illustrations). The darkest and lightest lice were used to create four pairs of dark lice and four pairs of light lice. Each pair was isolated in one of eight "common garden" containers and allowed to breed. The frequency of offspring in each color category was proportional to the shaded bars shown for each generation. Inseminated F1 females of the most extreme colors were used to breed F2 offspring, which were then allowed to mate with siblings to produce F3 offspring. F3 sib matings among the darkest or lightest individuals produced mostly dark or light F4 offspring. After Busvine (1946).

that live on birds with bills that are especially well adapted for preening lice (chapter 4).

Cryptic coloration has also been demonstrated within a single species of bird louse. The generalist feather louse *Quadraceps punctatus* is known from 27 species of *Larus* gulls (Price et al. 2003). Timmermann (1952) divided *Q. punctatus* into eight subspecies based on size, color, and other morphological characteristics. The gulls on which these subspecies occur vary in color from dark grey to mostly white (fig. 5.12a–c). Remarkably, the color of the *Q. punctatus* subspecies is almost perfectly correlated with the color of

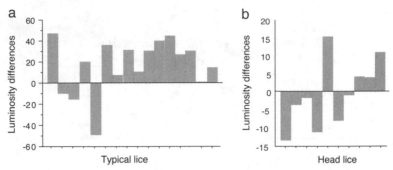

FIGURE 5.11. Differences in the luminosity scores of lice from related pairs of bird species with light or dark feathers. Each bar represents a single comparison of sister taxa. Positive values are cases in which the lighter-colored louse is found on the lighter colored host, and the darker louse is found on the darker host. Negative values are cases in which the lighter louse is on the darker host, or vice versa. (a) "Typical" lice, which are found on the body, wings, or tail of the host, showed significantly more positive than negative differences (P = 0.02). (b) "Head" lice, which are protected from preening, showed no association with host color (P = 0.68). Note the different scales on the y-axes of a and b.

their hosts (fig. 5.12c). This pattern is especially noteworthy given that most *Q. punctatus* subspecies occur on several species of gulls, many of which are not each other's closest relatives. For example, *Q. p. sublingulatus* is found on six species of *Larus* that are similar in color, yet not closely related (Pons et al. 2005). This pattern suggests convergent evolution of cryptic coloration among populations of single *Q. punctatus* subspecies.

Experimental evolution of crypsis in wing lice

Like *Larus* gulls, feral rock pigeons also vary dramatically in color, but within the single species *Columba livia*. Indeed, it is often possible to find pigeons varying from snow white to nearly black in a single flock. This variation is the legacy of artificial selection on color by pigeon breeders because, in most parts of the world, feral pigeons are descendants of escaped domesticated pigeons (Johnston and Janiga 1995; Shapiro and Domyan 2013). This variation in color makes it possible to study microevolutionary changes in the color of lice by transferring them among different-colored individuals of the normal host species.

Kim (2008) conducted a preliminary experiment of this kind using wild-caught rock pigeons and their wing lice. Lice were transferred from grey (wild type) rock pigeons to louse-free white pigeons and grey controls. Half of the birds were allowed to preen normally, while the remaining birds were impaired with bits. Over a period of nine months (about ten louse generations) the lice on white birds that could preen evolved lighter color-

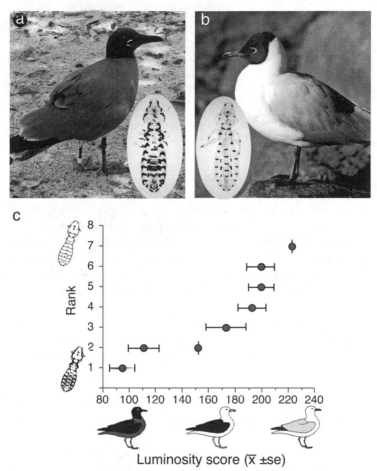

FIGURE 5.12. Variation in the color of gulls and subspecies of *Quadraceps punctatus*. (a) The lava gull (*Larus fuliginosus*) is host to a darker subspecies of louse (*Q. p. felix*) than (b) the black-headed gull (*Larus ridibundus*), which is host to a lighter subspecies (*Q. p. punctatus*); (c) Relationship between color ranks (1 = darkest) of eight subspecies of *Quadraceps punctatus* and the color of their hosts; e.g., the rank of *Q. p. felix* (a) = 1, whereas the rank of *Q. p. punctatus* (b) = 6. Note that two subspecies of lice have a tied ranking of 2. Circles with single vertical lines are subspecies of lice that parasitize a single gull species; most subspecies parasitize more than one species of gull. Gull photographs by (a) J. A. H. Koop and (b) F. Christriansen, (a and b) louse photographs courtesy of J.C. Stahl, Museum of New Zealand Te Papa, (c) from Bush et al. (2010).

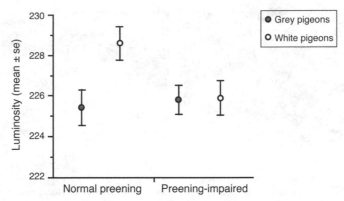

FIGURE 5.13. Experimental evolution of cryptic coloration in pigeon lice. Wing lice were transferred from grey (wild-type) rock pigeons to 12 (louse-free) white pigeons, and 12 (louse-free) grey controls. All birds were isolated in cages with one bird per cage. Half the birds could preen normally, while the remaining half had preening impaired with bits. After a period of nine months (ten generations of lice) the luminosity of approximately 12 lice from each bird was scored (methods as in Bush et al. 2010). (Lighter-colored lice had higher luminosity scores.) Lice on white birds that could preen were significantly lighter than lice on all other treatments, and there was a significant interaction between host color and preening ability (MANOVA, interaction P = 0.037) (Kim 2008 and unpublished data). Preening selected for the evolution of lighter coloration in the lice on white pigeons.

ation (fig. 5.13). The change in color was not because the lice fed on white feathers: lice on (bitted) white birds that could not preen did not become lighter in color.

Additional experiments confirmed that the color of lice is not influenced by the color of their food (box 5.2). The change in color was also unlikely to have been the result of sustained phenotypic selection within each generation of lice because the population sizes of lice on white and grey (preening) birds did not differ significantly (multivariate analysis of variance, effect of host color $P = 0.92$). It would be interesting to repeat this experiment with a common garden component to measure heritable variation in the color of the lice. It would also be interesting to explore the genomic architecture of cryptic coloration, analogous to work done on crypsis in vertebrates (Hoekstra 2010).

Counter-adaptations to host chemical defense

A few birds and mammals may use chemical defenses to combat lice and other arthropods (chapter 4). Most notably, pitohuis and their relatives in New Guinea have toxic feathers that can kill lice (Dumbacher 1999). Petri dish experiments show that, when given a choice, lice from other species of

BOX 5.2. You are not what you eat

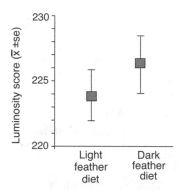

Most head lice on birds are darkly colored, regardless of the color of the host's plumage (Bush et al. 2010). This fact shows that the color of the feathers on which the lice feed does not dictate the color of the lice themselves (and see the data below). Moreover, the dark coloration of most avian head lice suggests that, in the absence of preening, dark coloration may be adaptive. Melanins, which are pigments that contribute to dark coloration, are thought to protect arthropods from the damaging effects of UV radiation (Majerus 1998; True 2003). For example, populations of *Daphnia longispina* inhabiting clear bodies of fresh water have more melanin than populations inhabiting murky water that blocks UV (Herbert and Emery 1990; Hobæk and Wolf 1991). The dark color of head lice could conceivably help to protect them against greater UV exposure on the head of the host.

To test whether diet influences the color of lice, 60 newly hatched wing lice (*C. columbae*) were removed from infested grey rock pigeons (*C. livia*) and placed on feathers pulled from black or white pigeons. The lice (with feathers) were kept in glass tubes in an incubator at optimal temperature and humidity (Nelson and Murray 1971) until they reached maturity, which takes about three weeks. Adult lice were then removed from the tubes, mounted on microscope slides, and scored digitally for luminosity (methods in Bush et al. 2010). Lice reared on the light versus dark feathers did not differ significantly in luminosity (see figure; ANOVA, df = 1, 19; P = 0.42). Thus, there was no evidence that the color of feathers influenced the color of lice feeding on them (unpublished data).

birds detect and avoid pitohui feathers in favor of nontoxic feathers of other species. Pitohui lice also live longer on pitohui feathers than do lice from nontoxic birds (although sample sizes for this result were very small). These preliminary data suggest that pitohui lice may have evolved some form of resistance to pitohui feather toxins (Dumbacher 1999).

It is well documented that lice have evolved counter-adaptations to chemicals used to combat lice on poultry, livestock, and humans (Levot 2000; Sangster 2001; Hodgdon et al. 2010). These chemicals include natural products, such as pyrethrins. Resistance to pyrethrins by human head lice, which in some cases is mediated by a single knockdown (*kdr*) point mutation (Hodgdon et al. 2010), is common in many regions of the world. By comparison, resistance is lacking in remote regions, such as Borneo, where people do not have access to head lice shampoos (Pollack et al. 1999). Resistance to pesticides by human lice shows that they have the capacity to evolve counter-adaptations to chemical defense.

Counter-adaptations to host immunity

Lice that feed on blood are presumably under selection to circumvent host immune defenses (chapter 4). Damage to the host's skin by feeding ectoparasites, such as fleas, as well as antigens in their saliva, are known to trigger immunological and hematological responses by the host. Counter-adaptations of ectoparasites include 1) skin-softening compounds that make penetration of host skin easier (Benjamini et al. 1961; Feingold and Benjamini 1961; Lehane 2005); 2) salivary compounds that delay the coagulation of blood; and 3) vasodilating compounds that increase blood flow, thus reducing feeding time (Lehane 2005).

As the host mounts an immune response, tissues swell dramatically, increasing the distance between the skin surface and blood vessels. In some cases, arthropods can no longer feed effectively because their mouthparts are too short to access blood over the increased distance (Owen et al. 2009). As a result, parasites are forced to feed on other regions on the host's body, where they may be more vulnerable to grooming. The immune response can also cause irritation that directs the attention of the grooming host to particular regions, thus providing an example of interaction between immunological and behavioral defenses (Lehane 2005; Schmid-Hempel 2011). Some blood-feeding ectoparasites, such as ticks, also secrete anti-inflammatory compounds in their saliva that delay the host's immune response, increasing foraging success and survival of the parasites (Ribeiro et al. 1985; Ribeiro and Francischetti 2003).

Almost no work has been done on counter-adaptations by lice to host immune defenses. What little we know is based on biochemical analyses of salivary proteins from human head and body lice (Jones 1998). The saliva of *P. humanus* contains vasodilators, anticoagulants, and immunoreactive proteins, only a few of which have been fully characterized (Mumcuoglu et al. 1996). Unlike most other blood-feeding insects, human lice inject but a small fraction of their salivary compounds at each feeding site. Jones (1998) observed feeding lice and repeatedly interrupted and moved them to new feeding locations. He estimated that less than 10% of the salivary stock was injected into the host at each feeding attempt, compared to 40% injected by *Anopheles albimanus* mosquitoes. By injecting only a small proportion of saliva at each feeding site, lice may be able to feed more quickly, then move to new locations to avoid host grooming in response to the itching caused by localized immune responses.

6   COMPETITION AND COADAPTATION

... the perpetual struggle for room and food.
—Malthus 1798

P arasites do not live in isolation, but are members of diverse parasite communities that share hosts. The most
thoroughly studied species on earth, *Homo sapiens*, has
more than 400 known species of parasites and pathogens
(Kuris 2012). Another well-studied species, the rock pigeon, has more than 70 known species of parasites (Johnston and Janiga
1995). While no individual harbors all parasites known from that host species, most individuals support more than one species of parasite at a time.
Sometimes the presence of a given parasite species facilitates infection by
other parasites through weakening of the host's behavioral, immunological,
or other defenses (Christensen et al. 1987; Stone and Roberts 1991; Krasnov
et al. 2005b; Fenton and Perkins 2010; Lafferty 2010; Telfer et al. 2010).

In other cases, the presence of a given parasite species can have a negative effect on other parasites due to competition for limiting resources in
or on the shared host (Dobson 1985; Esch et al. 1990; Poulin 2007; Goater
et al. 2014). Even parasites that exploit very different parts of a host's body
may compete, because each host individual is ultimately a single resource
(Janzen 1973). Interspecific competition can lead to the coadaptation of
traits that reduce the intensity of competition (Schluter 2000a,b, 2010; Pfennig and Pfennig 2012). Thus, in addition to coevolving with the host species,
parasites can also coevolve with other parasites that share that same host
species (fig. 6.1). The preceding chapters of this book have focused on interactions between hosts and parasites. This chapter focuses on interactions
between different species of parasites, and the role of the host in mediating
those interactions. We begin by considering competition between parasite
species, in general. We then provide a more detailed overview of competition between species of lice.

Interspecific competition in parasites
The first evidence for interspecific competition in natural populations of
parasites came from studies showing that coinfected hosts are less common than expected by chance. This pattern has been documented for
mammal helminths (worms) (Dobson 1985; Poulin 2001), digenean worms

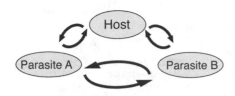

FIGURE 6.1. Parasites interact with their hosts, but they also interact with other parasite species that share those hosts. Arrows show possible reciprocal effects between a single host and two parasite species.

in snails (Kuris and Lafferty 1994), and parasitic copepods, tapeworms, and monogenean worms in fish (Paperna 1964). For example, carp (*Cyprinus carpioco*) infected with different species of monogeneans are relatively rare. Experiments manipulating monogenean (*Dactylogyrus* spp.) communities of captive fish confirm that this pattern is due to interspecific competition: for example, *D. extensus* is competitively excluded when the fish also have *D. vastator* (Paperna 1964).

Direct competition

In some cases, interspecific competition in parasites is direct (box 6.1). For example, some parasites compete by releasing toxins that kill or inhibit the growth or reproduction of competitors (Mideo 2009). Helminth worms secrete toxins that inhibit the growth of other species of worms (Cook and Roberts 1991). These toxins may be adaptations to facilitate competition with other worms, as well as more distantly related parasites (Roberts 2000). For example, Ferrari et al. (2009) demonstrated competition between gastrointestinal helminth worms (*Heligmosomoides polygyrus*) and ticks (*Ixodes ricinus*) coinfecting yellow-necked mice (*Apodemus flavicollis*). Tick populations suffered in the presence of worms, leading the authors to speculate that toxins released by the worms may suppress tick populations. Experiments are needed to further test this hypothesis (Ferrari et al. 2009).

Direct competition also occurs between species of ectoparasites. Krasnov et al. (2005a) investigated competitive interactions between flea species (*Xenopsylla conformis* and *X. ramesis*) co-infesting Sundevall's jird (*Meriones crassus*), a gerbil-like rodent. Their experiments revealed that, when the two species of fleas occur together, *X. conformis* populations are suppressed, particularly if food is limiting. Patterns of flea mortality led Krasnov et al. (2005a) to suggest that the mechanism for competitive suppression in this system is interspecific predation. The authors suggested that *X. ramesis* larvae prey upon *X. conformis* larvae. Interspecific predation occurs in other ectoparasites (Fox 1975), including lice (Rothschild and Clay 1957; Nelson and Murray 1971).

BOX 6.1. Competition and coadaptation

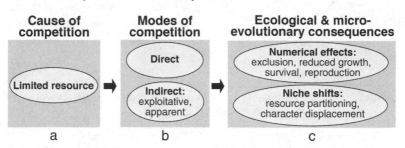

Cause of competition	Modes of competition	Ecological & micro-evolutionary consequences
Limited resource	Direct	**Numerical effects:** exclusion, reduced growth, survival, reproduction
	Indirect: exploitative, apparent	**Niche shifts:** resource partitioning, character displacement
a	b	c

Competition occurs when resources are limited (a). Interspecific competition is of two main types: direct and indirect (b; Thomson 1980). *Direct competition*, also known as antagonistic or interference competition, occurs when one species has direct numerical effects (c) on the members of another species. This occurs through physical combat, the secretion of toxins, or other mechanisms (Dobson 1985; Brown et al. 2009; Mideo 2009; Hatcher and Dunn 2011). *Indirect competition* occurs when one species has a negative effect on another species, but without any direct interaction. Two forms of indirect competition are exploitation and apparent competition.

Exploitation occurs when one species is better at acquiring some limiting resource than another species. The superior competitor may get to the resource faster, harvest it better, or make do with less of it. In some cases, exploitation is mediated by a third species. For example, predators can reduce competition between prey species by reducing prey population sizes (Paine 1966, 1974; Connell 1971). Similarly, hosts can reduce competition between parasite species by reducing parasite population sizes (Bush and Malenke 2008).

Apparent competition occurs when species compete for "enemy-free space," which provides protection from shared enemies, such as predators or parasitoids

Indirect competition

Competition between parasite species can also be indirect (box 6.1). Waage and Davies (1986) documented competition between different species of tabanid flies feeding on the blood of horses (*Equus ferus*). They found that the species *Hybomitra expollicata* and *Tabanus bromius* compete when they feed. Specifically, they found a negative correlation between the number of flies on a horse and the amount of time each fly spent feeding, even though the flies were not very dense. Neither food (blood) nor space (land-

(Holt 1977; Van Veen et al. 2006; Fenton and Perkins 2010; Cobey and Lipsitch 2013; Condon et al. 2014). For example, related species of leaf-mining flies in the tropics specialize on different host plants, preventing direct competition (Morris et al. 2004). Different species of flies share the same generalist parasitoid wasps. If the population size of one species of fly increases, triggering an increase in the parasitoids, it can lead to a decrease in the population sizes of the other species of flies, despite the fact that they do not affect one another directly. This is an example of apparent competition. Leaf miners that avoid detection by parasitoids, or kill parasitoid eggs with immune defenses, have the upper hand (Morris et al. 2004).

In some cases, interspecific competition can reduce survival or reproduction so much that it leads to the competitive exclusion of one species by another, resulting in local extinction of the competitively inferior species. From a geographic mosaic perspective (chapter 1), exclusion of one species creates a coevolutionary coldspot (Thompson 2005). Competition can be reduced through niche shifts that lead to better *resource partitioning* (c). For example, parasite species may partition resources by specializing on different regions of the host's body.

Selection to reduce competition can also lead to *character displacement* (c), which involves evolutionary changes in the interacting species. For example, competing species may evolve morphological or behavioral differences that help them partition a limiting resource. A well-known example is the evolution of different bill sizes among species of Darwin's finches that compete for overlapping food resources (Grant and Grant 2008). Character displacement is suspected when the distributions of species' traits diverge more in regions where the species co-occur than in regions where they do not co-occur (Taper and Case 1992; Schluter 2000a). Character displacement is an evolutionary response to reciprocal selective effects exerted between competing species. Character displacement is a type of coadaptation. It is also known as *coevolutionary displacement* (Thompson 2005).

ing sites) appeared to be limiting resources. However, the flies had to avoid host grooming (tail flicking) and, as the number of feeding flies increased, horses spent more time flicking their tails. This, in turn, left flies with less time to feed. In short, the flies were engaged in apparent competition (box 6.1) for feeding locations protected from host grooming.

Råberg et al. (2006) provided evidence for immune-mediated apparent competition between virulent and non-virulent clones of malaria (*Plasmodium chabaudi*) in mice. In immunocompetent mice, the avirulent clone suffered more from competition than did the virulent clone. In immunode-ficient mice, competitive suppression of the avirulent clone was alleviated and its relative density increased. The results of this study suggest that immune-mediated apparent competition may contribute to selection for increased virulence in mixed malaria infections.

Resource partitioning

Parasites can minimize or avoid competition by partitioning microhabitats in or on the host. Evidence for microhabitat partitioning is known from studies of tapeworm communities in skates and stingrays (Friggens and Brown 2005), as well as worms in cichlid fish (Vidal-Martínez and Kennedy 2000). These and other studies show that the degree of overlap between species of parasites is significantly less than that expected by chance. How-ever, these patterns could also be the result of noncompetitive microhabitat specialization, so they are not definitive evidence of resource partitioning due to competition. Note that monogenean gill parasites of fish and para-sitic flies of bats also exhibit microhabitat specialization; however, no evi-dence of competition has been found in either of these systems (Rohde 1979; Simková et al. 2000; Tello et al. 2008).

Comparison of hosts with single versus multiple infections of parasites can provide more conclusive tests of competition. Stock and Holmes (1988) investigated resource partitioning among species of intestinal worms in grebes, which support remarkably rich communities of worms. The authors collected 20 species of worms from western grebes (*Aechmophorus occiden-talis*), and 31 species of worms from eared grebes (*Podiceps nigricollis*). They found that different worm species were more or less restricted to specific regions of the intestine, with some overlap between species. In cases of coinfection, the worms showed significant reductions in microhabitat breadth along the intestine, consistent with resource partitioning (Stock and Holmes 1988).

As suggested by the work on grebe parasites, interspecific competition in parasites can involve more than two species. Lello et al. (2004) documented a network of competitive and facilitative interactions among several para-sites infecting wild rabbits. More recently, Condon et al. (2014), constructed an interaction network for a tri-trophic system consisting of two species of tropical squash (*Gurania* spp.) parasitized by 14 species of Tephritid fruit flies (*Blepharoneura* spp.). The flies, in turn, are parasitized by 18 species of parasitoid wasps (primarily *Bellopius* spp.). Complex interactions among the three trophic levels indicate that competition for host plants, as well as apparent competition to avoid and resist parasitoid attack, both contribute to the evolution and maintenance of fly diversity.

Interspecific competition in lice

Permanent ectoparasites, such as lice, provide an ideal opportunity to study competitive interactions because their microhabitat distributions and population sizes can be monitored closely over time on live hosts. Patterns of host use by lice led earlier workers to suspect that interspecific competition is important in lice (Hopkins 1949; Rothschild and Clay 1957). Hopkins (1949) noted that many mammals have just one species of louse. He speculated that competitive exclusion might drive this pattern. Anecdotal evidence supports the hypothesis that competition can lead to the local extinction of mammal lice.

The chewing louse *Trichodectes canis* is historically a parasite of dogs (Thompson 1940). Over the past century, *T. canis* has been displaced from much of its range by the chewing louse *Heterodoxus spiniger.* Prior to 1900, lice in the genus *Heterodoxus* were observed only on Australian marsupials, with the species *H. spiniger* occurring on the agile wallaby (*Macropus agilis*) (Thompson 1940). Over time, dogs have also become infested with *H. spiniger*, probably through contact with wallabies (Murray and Calaby 1971). Although *H. spiniger* now occurs on dogs throughout the world, there are few records of individual dogs infested with both *T. canis* and *H. spiniger* (Thompson 1940; Hopkins 1949; Barker 1994). This pattern suggests that *H. spiniger* may outcompete *T. canis*. However, experimental infestations are needed to test this hypothesis.

Co-infestation with two or more species of lice is common in some groups of mammals, such as hyraxes, civets, and rodents (Barker 1994; Oguge et al. 2009). The lice on these hosts may avoid competitive exclusion by partitioning microhabitats on the host. Reed et al. (2000b) examined the pelage of pocket gophers (*Thomomys bottae connectens*) co-infested with two species of chewing lice: *Geomydoecus aurei* and *Thomomydoecus minor.* They found that the two species tend to use different microhabitats, with *Thomomydoecus* found on more ventral regions of the host, and *Geomydoecus* found on more dorsal regions (fig. 6.2). These data are consistent with interspecific competition, although an experimental manipulation is needed for a more definitive test.

Lice on cattle also show evidence of microhabitat partitioning. Watson et al. (1997) examined cattle naturally infested with four species of lice. They found that each species of louse tends to use a different region of the host (fig. 6.3). Lewis et al. (1967) noted that when cattle are co-infested with two species of lice, the sucking louse *Linognathus vituli* is found primarily on ventral surfaces of the host, while the chewing louse *Bovicola bovis* is found primarily on dorsal surfaces. In single infestations on cattle restricted from grooming, however, the microhabitat preferences break down and the lice

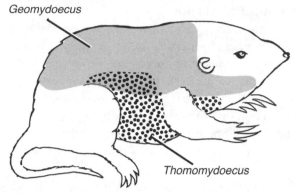

FIGURE 6.2. Distribution of lice on a pocket gopher (*Thomomys bottae connectens*). Shading indicates the relative distributions of two species of lice (*Geomydoecus aurei*, *Thomomydoecus minor*). This partitioning of microhabitats is consistent with interspecific competition (drawn by S.E.B., based on Reed et al. 2000b).

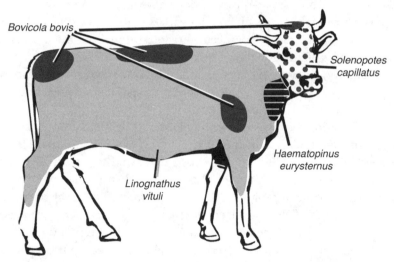

FIGURE 6.3. Distribution of lice on a steer. The preferred microhabitats of four species of lice are shown with different shading: *Bovicola bovis* on the poll (crown), back, rump, and shoulders; *Haematopinus eurysternus* on the dewlap; *Linognathus vituli* over most of the body; *Solenopotes capillatus* on the face, forehead, jaw, and muzzle. Drawn by S.E.B., based on Watson et al. (1997).

are more uniformly distributed over the host (Lewis et al. 1967). These data suggest that competition influences the microhabitat partitioning of at least two of the four species of lice infesting cattle.

Birds are usually co-infested by two or more species of lice (chapter 2). Compared to hair, feathers provide a more heterogeneous resource base that can be partitioned and exploited in different ways (Clay 1949b). Feather lice can be divided into four common ecomorphs that specialize on different microhabitats of the host: head lice, wing lice, body lice, and generalist lice (fig. 11.4) (Johnson et al. 2012). Birds are often co-infested with species of lice representing more than one of these different ecomorphs. Co-infestation with more than one species belonging to the same ecomorph is less common, although it does sometimes occur (plate 13). Clay (1949b) suggested that competition is responsible for the evolution of the different louse ecomorphs. We explore evidence in support of this hypothesis in chapter 11.

Lice also compete with other kinds of ectoparasites. Chen et al. (2011) tested for competitive interactions between lice (*Menacanthus stramineus*)

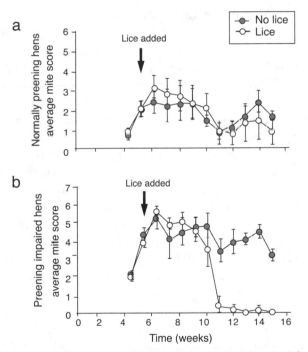

FIGURE 6.4. Host-mediated competition between lice and mites. Chen et al. (2011) experimentally infected chickens with northern fowl mites (*Ornithonyssus sylviarum*) and tracked their populations over time. Half of the hens (a) could preen normally, while the other half (b) had their preening impaired by trimming their beaks. Mite loads were allowed to establish and increase before lice (*Menacanthus stramineus*) were added to half of the birds in each treatment (arrows). Mite scores (mean ± se) are categorical rankings from 0 (no mites) to 7 (over 10,000 mites). Mite populations persisted significantly more often on normally preening birds with lice (P < 0.001). In contrast, mite populations on preening-impaired birds with lice went locally extinct. After Chen et al. (2011).

and northern fowl mites (*Ornithonyssus sylviarum*) on domestic chickens. They confirmed that the lice and mites compete, even to the point of competitive exclusion (fig. 6.4). In their experiments both preening and ectoparasite community composition were manipulated. Although competitive exclusion of mites by lice was common, the authors found that when birds could preen normally, the lice and mites coexisted. The results of this experiment suggest that host defense plays an important role in the maintenance of parasite species diversity. Hence, just as predatory starfish increase the diversity of prey species by preventing competitive exclusion (Paine 1966, 1974), host defense can prevent competitive exclusion in ectoparasites. This hypothesis has also been tested for the body and wing lice of pigeons, as we consider in the next section.

Competition between ecological replicate lice

Rock pigeons are often co-infested with body lice (*Campanulotes compar*) and wing lice (*Columbicola columbae*) (fig. 1.12). Bush and Malenke (2008) tested for competition between these two species of lice. The authors found no significant difference in the population sizes of wing lice (fig. 6.5a) or body lice (fig. 6.5b) on normally preening birds with single vs. mixed species infestations (although wing louse populations tended to be smaller if body lice were present). In contrast, wing louse populations were significantly lower in the presence of body lice (fig. 6.5a) on birds with impaired preening; body lice were not affected by the presence of wing lice (fig. 6.5b). Thus, body lice were competitively superior to wing lice, with host defense mediating their competitive effect.

What limiting resource is responsible for competition between body and wing lice? Bush and Malenke (2008) found that the two species of lice feed on the same feather barbules (fig. 6.6). Lice clip off the barbules with their mandibles, but even on heavily infested birds lice leave "stumps" behind, presumably because they are gape limited and the barbules are tapered in shape (fig. 6.6e). Barbules from birds infested with body lice only, or wing lice only, were grazed down to similar sized stumps (fig. 6.6f). These data show that wing lice and body lice feed on similar sized barbules, presumably because of similar gape limitations.

Bush and Malenke (2008) sampled feathers from eight plumage regions (fig. 3.4a), then compared the mass of feathers plucked from control birds without lice to that of feathers plucked from birds with wing lice only, or body lice only. Over the course of the 42-week experiment, feathers from infested birds weighed 24% less than feathers from uninfested birds. Wing and body lice did not consume feathers from all plumage regions equally: wing lice consumed more feathers from the back, nape, breast, keel, and

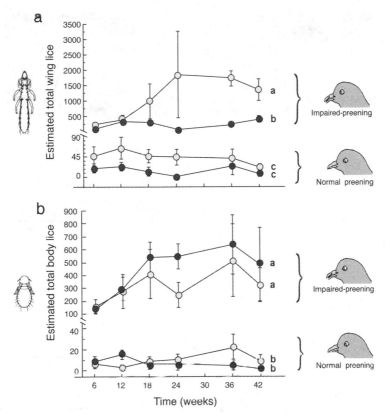

FIGURE 6.5. Population dynamics (mean ± se) of (a) wing lice and (b) body lice on rock pigeons with impaired-preening versus normal preening; open circles are birds with a single species of louse; black circles are birds with both species of lice. At the start of the experiment, birds were seeded with 100 wing lice, 100 body lice, or 50 wing lice + 50 body lice. Over the course of the experiment, lice were quantified repeatedly using visual censuses. Different lower-case letters indicate significant differences (P < 0.05). Comparisons among the eight treatments (= lines) show that 1) body lice competitively suppress wing lice, but only when host preening is impaired; 2) wing lice have no competitive effect on body lice. After Bush and Malenke (2008).

flank, whereas body lice consumed more feathers from the side, vent, and rump. Thus, there was evidence that wing and body lice concentrate their feeding on different body regions, to some extent. However, wing and body lice both consumed at least some feathers from all eight regions, demonstrating resource overlap and the potential for competition.

Wing and body lice both experienced resource partitioning in the face of competition, but the shifts exhibited by wing lice were more pronounced. Wing lice were found significantly less often on the torso of pigeons when

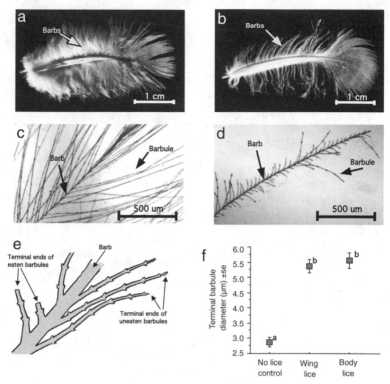

FIGURE 6.6. Consumption of feather barbules by wing and body lice. (a) Photo of a normal rump feather; (b) Photo of a rump feather severely eaten by lice; (c) Photo of feather barb with normal barbules; (d) Photo of feather barb with barbules severely eaten by lice; (e) Illustration of the differences in the terminal diameters of eaten and uneaten feather barbules; (f) Mean (± se) diameter of the terminal ends of uneaten barbules, compared to barbules consumed by wing lice (only) or body lice (only). Lower-case letters indicate significant differences. From Bush and Malenke (2008).

body lice were present (fig. 6.7a). Wing lice feed on the torso, so even though relatively few lice were seen there, this microhabitat is critical for their survival: shifts away from the food source may reduce the foraging efficiency of wing lice. In contrast, body lice did not shift their distribution at this scale; body lice were always found on the torso, regardless of the presence of wing lice (fig. 6.7b).

Microhabitat partitioning by both species of lice was also apparent at a finer spatial scale: among feather regions on the torso. Both species of lice were most commonly found on the rump feathers when the other species was absent (fig. 6.7c,d). This is not surprising because rump feathers weigh nearly twice as much as feathers from either the back or keel, thus providing the most food and space. When the other species of louse was present,

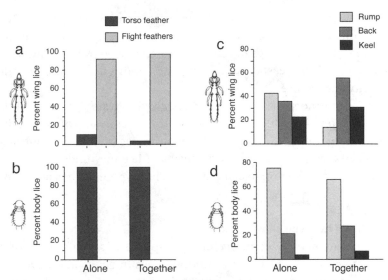

FIGURE 6.7. Microhabitat distributions of wing and body lice. Wing lice (a) shift microhabitat significantly in the presence of body lice (P < 0.0001), whereas body lice (b) remain exclusively on torso feathers, regardless of the presence of wing lice (P = 1.0). (Wing and body lice both feed on torso feathers.) The distribution of both wing lice (c) and body lice (d) among finer-scale regions on the torso is also influenced by the presence of the other species of louse. Although both wing and body lice show significant fine-scale shifts in the presence of the other louse (P < 0.01), the response of wing lice is greater than that of body lice. After Bush and Malenke (2008).

both wing and body lice shifted away from the rump to other regions of the torso. This shift was most dramatic for wing lice; 29% of wing lice were displaced from the rump in the presence of body lice (fig. 6.7c), compared to 11% of body lice displaced in the presence of wing lice (fig. 6.7d).

At the conclusion of the 42-week-long experiment, the birds still supported sizeable populations of lice (fig. 6.5). Thus, the lice did not exhaust all of the edible plumage on the birds (Bush and Malenke 2008). Resources need not be severely limiting for competition to take place. If use of the resource by one species merely reduces the efficiency of the second species, then the two species may compete (Thomson 1980). This is presumably the mechanism underlying competition for food between wing and body lice on pigeons.

Why are body lice competitively superior to wing lice? At least two factors may play a role. First, body lice are smaller and require less food than wing lice (Bush 2004). Persson (1985) predicted that, in cases of competition for food, the smaller species will be competitively superior. Second, body lice have the advantage of spending all of their time in the microhabitat where they feed. Wing lice must feed in the same microhabitat as body lice because flight feathers are too coarse to consume. However, wing lice must "commute" from the wings or tail to the torso to feed (Nelson and Murray 1971; Bush 2009). The cost of commuting may contribute to the competitive inferiority of wing lice.

Changes in morphology or other phenotypic traits frequently follow shifts in resource use (Schluter 2000a,b). The presence of competition between wing and body lice, and the effect it has on the distributions of both species, suggest that the two species of lice may undergo character displacement in response to competition. This hypothesis has not yet been tested.

Competition between congeneric wing lice

Some birds are parasitized by two or more species of a single genus of lice. For example, of 157 species of Columbiformes (pigeons and doves) with known species of *Columbicola*, 34 (22%) also host at least one other species of *Columbicola* (Bush et al. 2009b). Given their similarity, coexisting species of *Columbicola* might be expected to compete. Experiments with *C. bacuoides* and *C. macrourae*, both of which parasitize mourning doves (*Zenaida macroura*), provide support for this hypothesis (Johnson et al. 2009; Malenke et al. 2011). Despite the competitive effect of *C. macrourae* on *C. baculoides* (fig. 6.8), the two species of lice coexist on mourning doves under natural conditions. Coexistence may be explained by condition-dependent competition.

Condition-dependent competition is a form of resource partitioning mediated by variation in environmental conditions, such as differences in pH, temperature, or humidity. Inferior competitors that can withstand harsh environmental conditions can minimize or avoid competition by exploiting environments that are unsuitable for superior competitors (Dunson and Travis 1994; Parsons 1996; Greenslade 1983). For example, of the two *Columbicola* species on mourning doves, *C. baculoides* is found mainly in arid western North America, while *C. macrourae* is found mainly in the more humid east (plate 14; Malenke et al. 2011).

Most lice cannot tolerate arid conditions (Rudolph 1983; Moyer et al. 2002a), and this is the case with *C. macrourae*. In contrast, *C. baculoides* can tolerate low humidity (fig. 6.9) (Malenke et al. 2011). Under experimental conditions of high humidity, *C. macrourae* outcompetes *C. baculoides* (fig. 6.8). These experimental results suggest that *C. macrourae* outcompetes *C. baculoides* in the humid east, while *C. baculoides* persists in the arid west under conditions that *C. macrourae* cannot tolerate. In short, arid conditions essentially provide *C. baculoides* with a refuge from competition.

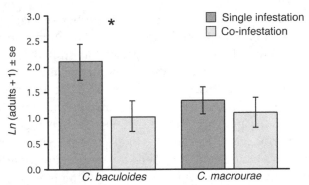

FIGURE 6.8. Populations of *Columbicola baculoides* and *C. macrourae* on mourning doves. Birds were seeded with a single louse species, or co-infested with both species. Each bar is the mean number of lice on 11 birds. *C. baculoides* populations are significantly larger when they are alone, compared to when they share a host with *C. macrourae* (*P < 0.05). In contrast, *C. macrourae* is unaffected by the presence of *C. baculoides*. After Malenke et al. (2011).

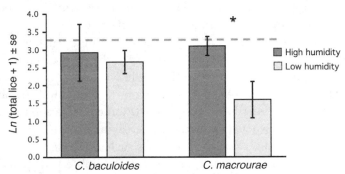

FIGURE 6.9. *Columbicola baculoides* and *C. macrourae* populations on mourning doves after 6 weeks' exposure to high-humidity (~65% relative humidity) and low-humidity (~45% relative humidity) conditions. Each bar is the mean number of lice on six birds. The dotted line shows the size of the experimentally transferred populations (n = 25 lice per bird). *Columbicola baculoides* was not significantly affected by humidity, whereas *C. macrourae* populations were significantly smaller at low humidity (*P = 0.05). After Malenke et al. (2011).

Studies of global climate change predict that much of North America will become more arid over the next century (International Panel on Climate Change 2007). If this happens, then *C. baculoides* may undergo a range expansion, with *C. macrourae* becoming extirpated from regions in which it currently occurs (Malenke et al. 2011). Similar changes have been documented for other parasites. For example, two tick species that infest Australian reptiles have a parapatric distribution, with one species found in arid regions (*Ambylomma limbatum*) and another species (*Aponomma hydrosauri*) found in more humid regions (Bull and Burzacott 2001). The geographic boundary between these species has shifted as climate has changed. Condition-dependent competition may be a factor driving the geographic shifts in these tick species (Bull and Burzacott 2001).

Models predicting the co-extinction of parasites with their hosts have not incorporated the influence of condition-dependent competition among parasites (Koh et al. 2004; Dunn et al. 2009). Current estimates of parasite coextinction rates may therefore underestimate parasite loss, since parasites experiencing condition-dependent competition may go extinct long before their hosts show any evidence of decline.

Competition-colonization trade-offs in wing lice

Competition-colonization trade-offs can facilitate the coexistence of species in free-living systems (Levins and Culver 1971; Tilman 1994). In such cases, the inferior competitor persists if it is better than the superior competitor at colonizing unexploited resources. Thus, inferior competitors should be superior dispersers. Competition-colonization trade-offs have been documented in microbes, plants, and insects (Amarasekare 2003); however, there is some controversy about the pervasiveness of the phenomenon. Models indicate that a competition-colonization trade-off is likely to occur between species where dispersal is limited and resources are patchily distributed (Yu and Wilson 2001). Predictions from competition-colonization models can be tested in systems where the relative dispersal rates of competing species can be measured and manipulated. Lice are such a system.

Feather lice on pigeons and doves are dispersal limited and patchily distributed. If inferior competitors can escape competition by moving to uninfested host individuals, then dispersal could be selectively advantageous. Two predictions concerning the relative dispersal abilities of the feather lice on pigeons and doves follow from competition-colonization trade-off models. First, competition between congeneric species of wing lice should lead to the increased dispersal of the species that is the inferior competitor. Second, competition between wing and body lice should lead to the increased dispersal of wing lice, given that they are the inferior competitors.

Johnson et al. (2009) provided evidence in support of both predictions. The authors investigated potential competition-colonization trade-offs within *Columbicola* using a comparative phylogenetic approach. They evaluated the direction of changes in host specificity within the genus *Columbicola*, and examined how those changes are related to potential competitors (plate 13). The authors showed that *Columbicola* are significantly more likely to be found on more than one host species when congeneric species are present. In other words, they found that the evolution of "generalists" (lice that use more than one host species) is significantly correlated with the presence of competitors.

These data differ from the traditionally held view that competition promotes the evolution of parasite specialization by selecting for an increasing level of resource partitioning among competitors (Holmes 1973; Futuyma and Moreno 1988; Poulin 2007). The data also challenge the once commonly held view that specialization is an evolutionary dead end (Simpson 1944; Colles et el. 2009, but see Nosil 2002; Nosil and Mooers 2005; Forister et al. 2012). Experimental studies are needed to test whether generalist *Columbicola* species are, in fact, less competitive than specialist *Columbicola* species. The competition-colonization hypothesis further predicts that wing lice, which are inferior competitors to body lice, are better at dispersing than body lice, despite their similar life histories. We review the dispersal ecology of lice in the next chapter, including the comparative dispersal ability of wing and body lice.

III HOSTS AS ISLANDS

7 DISPERSAL

Almost every anomaly in the distribution of animals and
plants may be explained by a careful consideration of the
various means of dispersal which organisms possess.
—Wallace 1893

Dispersal is the movement of individuals away from
their place of birth (Levin 2009; Krebs 2009). Dispersal
governs gene flow and thus population genetic structure. It is therefore of central importance to ecology
and evolution (Clobert et al. 2001, 2012). Selection and
dispersal are the two most important processes influencing the geographic
mosaic of coevolution (Thompson 2005). Through its effects on gene flow,
dispersal brings different genotypes of hosts and parasites into new combinations, and selection molds the outcome of these interactions (Forde et al.
2004; Morgan et al. 2005; Vogwill et al. 2010). Local adaptation of parasites
is governed by the dispersal rates of parasites in relation to those of the
hosts (Greischar and Koskella 2007; Hoeksema and Forde 2008). Limited
dispersal contributes to population subdivision, divergence, and ultimately
the process of diversification. Hence, data on dispersal are required to interpret coevolutionary dynamics in both micro- and macroevolutionary time
(Boulinier et al. 2001).

Parasite dispersal is often tied to host dispersal; when a host disperses,
so do its parasites (Blouin et al. 1995; Criscione and Blouin 2004). However,
parasite movement is not necessarily dependent on that of the host. Most
parasites also have dispersal mechanisms that are independent of the host.
Moreover, some parasites reciprocally influence host dispersal (box 7.1).
Parasite dispersal occurs over three main spatial scales: between individuals in a host population (fig. 7.1a), between host populations (fig. 7.1b), and
between host species (fig. 7.1c). Parasites occasionally disperse to novel host
species on which they are not normally found. If these parasites can establish viable breeding populations on the novel host, then they may undergo
a host switch.

Parasite ecologists use the term "transmission" when referring to movement of parasites between hosts of the same species, or between hosts of
different species if they are integral to completion of the parasite's life cycle.
Transmission is a form of dispersal and the terms are largely interchange-
able. From the point of view of the parasite, transmission tends to empha-

BOX 7.1. Effects of parasites on host dispersal

Parasite dispersal is fundamentally affected by host dispersal. The reciprocal can also be true; host dispersal can be affected by parasite dispersal. Although the potential influence of parasites on host dispersal is an understudied topic (Boulinier et al. 2001), there are many reasons why parasites might exert selection on host dispersal. Parasites may manipulate host dispersal to facilitate their own transmission to new hosts (Lion et al. 2006). However, in cases where parasites are locally adapted to hosts, they may manipulate hosts into dispersing shorter distances. Parasites have been shown to select against migrant hosts, effectively reinforcing the geographic diversification of hosts (MacColl and Chapman 2010). Moreover, hosts may be under selection not to disperse if their defenses are adapted to the local parasite community.

In one of the few experiments testing the influence of parasites on host dispersal, great tit (*Parus major*) fledglings from nests in which fleas were eradicated dispersed greater distances than fledglings from infested nests (Tschirren et al. 2007). Fledglings from infested nests had higher reproductive success when they dispersed shorter distances, but fledglings from parasite-free nests had higher reproductive success when they dispersed farther. Little is known regarding potential effects of lice on the dispersal or migration of bird or mammal hosts.

---

size movement "towards" a host, while dispersal emphasizes movement "away" from the host, to reduce inbreeding, for example (Bush et al. 2001). Transmission also tends to emphasize movement within a parasite population, whereas dispersal tends to emphasize movement between parasite populations.

In this chapter we review three major modes of dispersal by lice. For simplicity, we refer to all three modes as forms of transmission: 1) vertical transmission (movement of lice directly from parent hosts to their offspring); 2) direct horizontal transmission (movement of lice between individual hosts that are in direct contact); and 3) indirect horizontal transmission (movement of lice between individual hosts that are not in direct contact). By definition, vertical transmission occurs between members of the same host species. The second and third modes of transmission can occur between members of the same or different host species.

Vertical transmission by lice

Because lice are permanent parasites that pass all stages of their life cycle on the body of the host, it should come as no surprise that their most com-

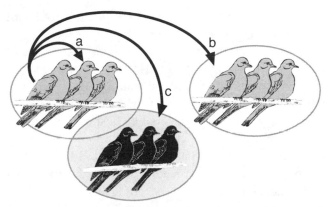

FIGURE 7.1. Spatial scales of parasite dispersal: (a) between host individuals in the same population, (b) between host populations, (c) between host species. Arrows indicate parasite dispersal.

mon mode of dispersal is vertical transmission. Vertical transmission has been measured for lice from a diverse assemblage of mammals, including squirrels, woodrats, sheep, fur seals, and others (Linsdale and Tevis 1951; Holdenried et al. 1951; Murray 1963; Kim 1972, 1975; Marshall 1981a). Vertical transmission has also been measured for lice from a diverse assemblage of birds, including seabirds, pigeons, swifts, blackbirds, bee-eaters, and others (Eveleigh and Threlfall 1976; Marshall 1981a; Clayton and Tompkins 1994; Lee and Clayton 1995; Darolova et al. 2001; Brooke 2010).

To enhance the probability of successful vertical transmission, some lice coordinate their reproductive efforts to coincide with those of the host. A dramatic example is the amphibious lice parasitizing marine mammals, such as seals and sea lions. The fur of some marine mammals protects lice and their eggs (Murray and Nicholls 1965); however, the fur of South American sea lions (*Otaria flavescens*) is too sparse to protect the eggs of the sucking louse *Antarctophthirus microchir* (Aznar et al. 2009). Reproduction by this louse is limited to periods when sea lions haul out on land each summer to bear young. Female sea lions have their pups, then alternate between nursing on land and foraging at sea. This means the pups are the only hosts remaining on land long enough for the lice to breed. Most pups are infested with lice the day they are born; the pups then remain on land for about 30 days (Aznar et al. 2009; Leonardi et al. 2013). The louse *A. microchir* has one of the shortest generation times known for lice, at about two weeks. This generation time is so short, in fact, that that the lice can complete two generations within the brief time that sea lion pups spend on land (Aznar et al. 2009).

Life cycles of some bird lice may also be synchronized with host life cycles (Marshall 1981a). Foster (1969) reported an increase in reproduction of lice on orange-crowned warblers (*Vermivora celata*) that coincided with the host's reproductive season. Foster found that eggs of amblyceran (blood-feeding) lice were most common on museum skins of warblers collected immediately prior to the breeding season. The author suggested that breeding by the lice could have been triggered by host reproductive hormones, similar to the demonstrated influence of host hormones on reproduction in rabbit fleas (Rothschild and Ford 1964). Consistent with this hypothesis was the fact that ischnoceran (feather-feeding) lice on the same birds showed no population increases prior to host breeding, perhaps because they were not exposed to hormones in the host's blood. Additional work is needed to test the hypothesis that host hormones trigger reproduction in bird lice.

Vertical transmission has also been measured directly in nesting birds. Lee and Clayton (1995) monitored the transmission dynamics of *Dennyus hirundinis*, an amblyceran louse that parasitizes common swifts (*Apus apus*) in Europe. When not on the nest, swifts spend all of their time flying. Remarkably, they even sleep on the wing (Rattenborg 2006; Liechti et al. 2013). Thus, there is little opportunity for *D. hirundinis* to disperse between hosts except during the breeding season. The main mode of dispersal is vertical transmission between parents and offspring at the nest. Lice on parent birds increase in number while the nest contains young, naked nestlings. Once the nestlings' feathers emerge—at about two weeks of age—lice nymphs move onto them and mature into adult lice. New adult females begin laying eggs on nestlings before the birds fledge from the nest (fig. 7.2).

In a study of lice on breeding blackbirds (*Turdus merula*) Brooke (2010) showed that vertically transmitted lice aggregate on the largest nestlings in each brood, possibly because they are the first ones to develop feathers. Marshall (1981a) reviews additional examples of vertical transmission, as well as other modes of dispersal by both bird lice and mammal lice.

Direct horizontal transmission by lice

Lice do not rely solely on vertical transmission. Direct horizontal transmission is also common, especially between mates and other hosts that come into direct contact while roosting, nesting, or otherwise interacting. Darolova et al. (2001) investigated the transmission of *Brueelia apiastri* between European bee-eaters (*Merops apiaster*). The authors found that even though most adult birds are infested with lice (98.3% of 176 adults), few nestlings are infested prior to fledging (10.8% of 167 nestlings). The authors noticed that the prevalence of *B. apiastri* on paired male and female birds was correlated, suggesting horizontal transmission between mates.

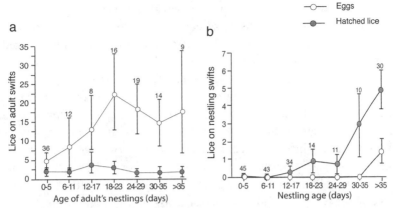

FIGURE 7.2. Transmission of lice (*Dennyus hirundinus*) from adult to juvenile common swifts (*Apus apus*) over the nesting cycle. (a) Plot of the number of lice and eggs on adult birds in relation to age of their nestlings. (b) Plot of the number of lice and eggs on nestling birds of different ages. Nestlings were infested with lice (from their parents) as early as two weeks, by which time the tips of their feathers had emerged. Error bars are 95% bootstrapped confidence intervals. Numbers above bars are the number of birds examined during each interval. The data include repeat examinations of the same birds between intervals, but no repeats within intervals. After Lee and Clayton (1995).

To test this hypothesis, they exterminated lice on one member of each of eight breeding pairs. When the birds were examined again five weeks later, all eight parasite-free birds were reinfested with lice, indicating probable horizontal transmission. Thus, even in cases where rates of vertical transmission are low, horizontal transmission appears to be common enough to maintain a high prevalence of lice among adult bee-eaters.

Some lice have behavior that appears to enhance direct horizontal transmission. For example, lice on ring-necked pheasants (*Phasianus colchicus*) congregate on the backs of female birds during the breeding season, facilitating transmission to males during copulation, which lasts only a few seconds. Hillgarth (1996) placed sticky tape on the legs of lice-free male pheasants, then examined the tape after the males had copulated with louse-infested females. Five of 20 male pheasants had lice on their tape.

Generally speaking, mammal lice have more time to disperse between mating hosts because copulation lasts longer in mammals than birds. *Pthirus pubis*, which moves horizontally during host copulation (Buxton 1947), has a window of opportunity that averages about seven minutes (Corty and Guardiani 2008). Lice also sometimes rise to the surface of fur when one host rubs against another one (Matthysse 1946; Murray 1963); this behavior may enhance transmission (Durden, pers. comm.). Lice on rodents crawl out to the tips of hairs, which may also increase the probability of transmission from a dead or dying host to a healthy one (Westrom and Yescott 1975).

The high frequency of direct horizontal transmission was illustrated by a mark-release-recapture experiment carried out by Harbison et al. (2008). Seventy feral rock pigeons were trapped from a population of 350–400 individuals under a large highway overpass. The birds were taken into captivity housed at low humidity for several weeks to clear them of "background" lice and viable eggs (Moyer et al. 2002a). After verifying that the birds were completely free of lice, they were released back at the bridge where they were captured. Of 32 birds the authors managed to recapture less than one month later (within a louse generation), 94% were reinfested with lice. The only source of lice was direct horizontal transmission from other infested birds. No parasitic flies were found at the bridge, meaning that indirect horizontal transmission of lice by phoresis could not have been a factor (see below).

Horizontal transmission is an essential means of dispersal for the lice of brood parasites, such as cuckoos, cowbirds, and indigobirds, which dupe other bird species into rearing their offspring (fig. 7.3). Since brood parasites have no contact with their own nestlings, it is not possible for their lice to be vertically transmitted. Lice from foster parents do get vertically transmitted to brood parasitic nestlings, but they do not establish viable breeding populations (Clayton and Johnson 2001). Some brood parasites, such as cuckoos, have host-specific genera of lice that are found only on them (Price et al. 2003; Balakrishnan and Sorenson 2006). How these lice get transmitted to the next generation of brood parasites has long fascinated students of ectoparasites and brood parasites alike.

Brooke and Nakamura (1998) shed light on this question for the lice of common cuckoos (*Cuculus canorus*). The authors determined that first-year birds must acquire their lice between leaving the nest in summer and returning from migration the following spring. Of 21 cuckoo nestlings examined, none had cuckoo lice. Nor were any cuckoo lice found in 19 nests containing cuckoo eggs, ruling out the unlikely possibility that cuckoo lice disperse from adult cuckoos to the nests of foster parents and wait for the cuckoo eggs to hatch. Cuckoos making their first return to breeding grounds from the wintering grounds had just as many lice as older birds. The authors concluded that the most likely transmission route for cuckoo lice is horizontally from adult to juvenile birds during feeding aggregations on the wintering grounds.

As in the case of bird lice, most work on the transmission biology of mammal lice has inferred transmission by comparing the prevalence or intensity of lice among host individuals (as in fig. 7.2). However, two mark-release-recapture studies have quantified lice transmission more directly.

FIGURE 7.3. Common cuckoo (*Cuculus canorus*) chick ejecting a foster reed warbler's (*Acrocephalus scirpaceus*) egg from the nest. A reed warbler nestling has already been ejected. The parent warbler watches, but takes no preventive action (reprinted from Davies 2000, *Cuckoos, Cowbirds and Other Cheats*, by permission of Poyser and Bloomsbury Publishing Plc.).

Durden (1983) clipped setal hairs of lice (*Hoplopluera erratica*) on Eastern chipmunks (*Tamias striatus*), with all lice from an individual host receiving an identical clipping pattern. Marked lice were returned to their chipmunk hosts, which were immediately released at the capture site. The chipmunks were recaptured periodically and their lice identified. Over the course of the six-month study, 66.7% of marked lice moved to different host individuals, mostly in the summer breeding season. Horizontal transmission between opposite sex hosts was common (27 cases), but some transmission of lice between same-sex hosts also occurred (6 cases). Male-male aggression was presumed to be the mechanism for exchanges between same-sex hosts. Overall, male hosts were the most frequent donors of lice over the course of the study.

A recent study of color marked lice (*Lemurpediculus verruculosus*) on brown mouse lemurs (*Microcebus rufus*) in Madagascar documented lice

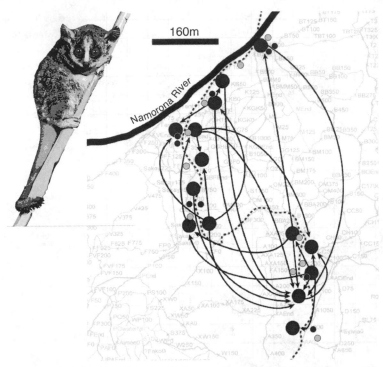

FIGURE 7.4. Map of louse transfers recorded among lemurs. The dotted line shows the 1-km trapping transect; circles represent the average trap locality for each individual lemur, with black circles representing males and grey circles representing females. Arrows connecting large circles indicate at least one recorded transfer between individual lemurs. Small circles indicate lemurs that neither received nor donated any lice. None of the female lemurs received or donated lice. After Zohdy et al. (2012).

dispersing exclusively between male lemurs (fig. 7.4; Zohdy et al. 2012). The authors suggested that the lice moved between hosts during bouts of aggression between males in the breeding season.

Lice can also move by direct horizontal transmission between hosts of different species. Interactions between birds in mixed-species foraging flocks may provide one opportunity for transmission between host species, although this has not been carefully studied (Freed et al. 2008a). Another mechanism involves the "straggling" of lice from prey to predator species (Ansari 1947; Palma and Jensen 2005; Bush et al. 2012). For example, Whiteman et al. (2004) documented two species of lice from Galápagos doves (*Zenaida galapagoensis*) on Galápagos hawks (*Buteo galapagoensis*), which prey upon the doves. Molecular analyses confirmed that the lice on hawks came from doves. Persistent transmission of lice to predators can lead to more regular host sharing by lice. Perhaps the best example is *Colpocephalum tui binatum*, which parasitizes several species of pigeons and doves, and several species of raptors that feed on them, such as the peregrine falcon (*Falco peregrinus*). Direct horizontal transmission is almost certainly responsible for this species of louse being found on unrelated host orders (Price et al. 2003).

Indirect horizontal transmission by lice

Lice also use indirect horizontal transmission to disperse between hosts of the same or different species. Clay (1949b) and Timm (1983) suggested three main mechanisms: 1) shared nests or nest holes, 2) shared dust baths, or 3) detached feathers. Competition for nest holes can be intense, with superior competitors taking nest holes or nesting material away from inferior competitors (Merilä and Wiggin 1995; Fey et al. 1997). Lice are sometimes observed in nests (Nordberg 1936), which may serve as fomites enhancing the indirect transmission of lice between birds of the same or different species. Shared nests and nest holes explain patterns of host sharing in owl lice (Clayton 1990a), toucan lice (Weckstein 2004), and songbird lice (Johnson et al. 2002a).

Shared dust baths are another way in which lice may disperse indirectly between hosts. Hoyle (1938) used a series of simple experiments to show that chicken lice can disperse to house sparrows (*Passer domesticus*) by way of shared dust baths. Dust bathing is a possible explanation for the fact that different species of ground-dwelling doves share more lice than different species of arboreal doves (Johnson et al. 2011a). Note that not all types of dust are effective at killing lice (chapter 4). Therefore, it may be the case that lice dislodged during bouts of dust bathing are capable of surviving on new hosts that use the same dust-bathing site. However, this requires further testing.

Live lice have also been observed on molted feathers of waterfowl (Eichler 1963). If floating feathers come into contact with swimming or wading birds, or if the feathers are used for nesting material, this may be another mechanism for horizontal dispersal of lice between the same or different host species. Walking on water is yet another possible mechanism for the indirect transmission of lice on aquatic birds. Members of the amblyceran genus *Trinoton*, which parasitize waterfowl, have tarsal setae that enable them to walk or run quickly across the surface tension of water (Stone 1969; Gustafsson and Campbell, pers. comm.). When individuals of *Trinoton querquedulae* were placed in a small pool, they quickly oriented toward ripples in the water caused by swimming ducks, or a finger moving in the water (the lice did not move towards stationary objects in the water). This behavior deserves further investigation. It may well be a common mode of indirect

Table 7.1. Distribution of lice on hippoboscid flies (*Ornithomya fringillina*) collected from six species of birds in a field study on Fair Isle, Shetland, UK (Corbet 1956b)

Bird species	# birds	# flies	% flies with lice	# phoretic lice
Starling (*Sturnus vulgaris*)	68	156	43.6%	164 lice on 68 flies
Wheatear (*Oenanthe oenanthe*)	114	243	0.4%	1 louse on 1 fly
Rock pipit (*Anthus spinoletta*)	113	208	1.0%	5 lice on 2 flies
Meadow pipit (*Anthus pratensis*)	41	67	1.5%	1 louse on 1 fly
Twite (*Carduelis flavirostris*)	14	23	0%	0
Merlin (*Falco columbarius*)	NA	35	2.9%	6 lice on 1 fly

NOTE. All lice were *Sturnidoecus sturni*, a host-specific parasite of starlings (Price et al. 2003). The data show that even host-specific lice have opportunities to disperse among species of birds via phoresis. The data also show that phoresis can be common and lead to the dispersal of clusters of lice. For example, in one case 31 phoretic lice were removed from a single fly!

horizontal transmission between waterfowl of different species. Indeed, it could explain why *T. querquedulae* has been collected from 68 different species of waterfowl (Price et al. 2003)!

Perhaps the most intriguing way in which lice disperse indirectly between hosts is by hitchhiking on other parasites, such as hippoboscid flies (plate 5). This behavior, which is known as phoresis, allows lice to disperse to new individuals of the same or different host species (Harbison and Clayton 2011). Corbet (1956a,b) used mark-release-recapture methods to demonstrate that *Sturnidoecus sturni*, which normally parasitizes European starlings (*Sturnus vulgaris*), is frequently phoretic on hippoboscid flies (table 7.1). Phoretic lice use their mandibles or claws to grasp the setae or cuticle of flies (Keirans 1975a,b; Marshall 1981a). Since the flies are not as host specific as lice, they provide a mechanism for the dispersal of lice among host species. Harbison and Clayton (2011) confirmed experimentally that this can occur in the case of pigeon wing lice (see next section). Species of lice vary in their ability to remain attached to flies (Harbison et al. 2009). Lice in the closely related genera *Brueelia* and *Sturnidoecus* are commonly recorded as phoretic, as are the wing lice of pigeons (*C. columbae*) (plate 5b; Keirans 1975a).

There are almost no published records of phoresis by amblyceran lice. It has been suggested that this is because of differences in amblyceran mandibles that make it difficult for them to hold onto flies (Keirans 1975a). However, sucking lice, which have no mandibles, are also phoretic (Durden 1990). Kirk-Spriggs and Mey (2014) reviewed all known cases of phoresy by sucking lice on flies. The most dramatic example was Mitzmain's (1912) observation of 620 sucking lice (*Haematopinus bituberculatus*) dispersing phoretically among 1,800 *Lyperosia* buffalo flies (Muscidae) that were examined. Most of the lice were nymphs, which used their legs to hang onto the legs of flies, suggesting that the claws of adult lice may be too large to grasp the fine legs of flies (Durden 1990).

Dispersal of ecological replicate lice

Pigeon and dove body and wing lice disperse using both vertical and horizontal transmission. As for the swifts discussed earlier (fig. 7.2), vertical transmission occurs once the tips of the nestlings' feathers emerge (Clayton and Tompkins 1994). The rate of vertical transmission of wing lice is higher than that of body lice. Harbison et al. (2008) showed that the mean number of wing lice on nestlings aged 3–4 weeks was 42% of that on parents, while the mean number of body lice on nestlings was only 21% of that on parents (fig. 7.5a). The prevalence of lice on young birds at the time of fledging was also higher for wing lice (93%) than body lice (71%) (Harbison et al. 2008).

By comparison, rates of direct horizontal transmission by wing and body lice are similar to each other (fig. 7.5b; Harbison et al. 2008). Of 32 birds cleared of lice, then released and eventually recaptured, 44% had wing and body lice, 25% had wing lice only, and 25% had body lice only. In contrast, wing and body lice differ dramatically in rates of indirect horizontal transmission. Wing lice hitch rides on hippoboscid flies, dispersing between individuals of the same (fig. 7.6a) or different (fig. 7.6b) host species. Body lice are not phoretic and thus do not disperse between hosts on flies (fig. 7.6a,b).

The reason for the dramatic difference in phoresis by wing and body lice is related to differences in morphology and behavior. Harbison et al. (2009) showed that wing lice are attracted to flies, whereas body lice are not. When wing and body lice are experimentally "forced" to attach to flies, wing lice are also better at remaining attached to active flies than body lice (fig. 7.7).

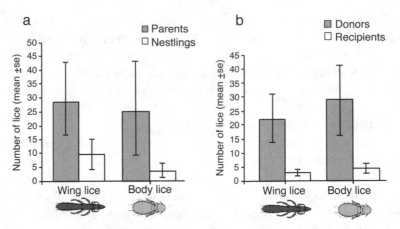

FIGURE 7.5. (a) Vertical transmission of lice from parent to nestling rock pigeons (n = 15 nests); transmission of wing lice was significantly greater than that of body lice (Wilcoxon signed-ranks test, P = 0.013). (b) Direct horizontal transmission of adult lice between wild birds (n = 30 donors, 32 recipients). There was no significant difference in horizontal transmission by wing versus body lice to parasite-free birds (open bars), relative to the number of lice on "donor" birds (dark bars). Bars are means with 95% confidence intervals. After Harbison et al. (2008).

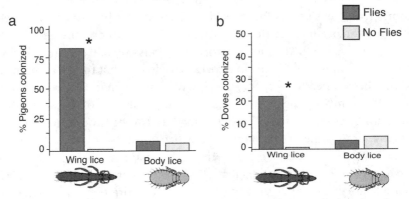

FIGURE 7.6. Phoretic transmission of wing and body lice to recipient rock pigeons and mourning doves. The experiment used individually caged birds housed in wooden sheds. On one side of each shed was a row of 20 louse-infested "donor" rock pigeons. On the opposite side of the shed was a row of louse-free, "recipient" pigeons (n = 10) interspersed with louse-free "recipient" doves (n = 10). Hippoboscid flies were released in experimental sheds, but not in the control sheds. Over the course of six-month experimental trials, wing lice colonized recipient pigeons (a) and doves (b) in sheds with flies, but not in the sheds without flies. Body lice were not phoretic; the low numbers of body lice on recipient birds, which did not differ significantly between experimental and control sheds, may have been due to contamination. Note different scales on y axes for a and b. * P < 0.05. After Harbison et al. (2008) and Harbison and Clayton (2011).

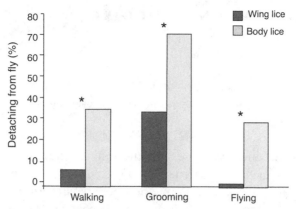

FIGURE 7.7. Detachment rates of wing and body lice from flies that were walking a distance of 5 cm, grooming for 10 minutes, or flying a distance of 3 meters. *P < 0.005. After Harbison et al. (2009).

The reason for the difference in attachment may have to do with the short legs of body lice, compared to the "outrigger" legs of wing lice (plate 5b). The evolution of legs for locomotion on large coarse wing feathers may have preadapted wing lice for phoretic dispersal.

Why do wing lice engage in phoresis, which is a risky behavior? Phoretic lice can fall off the fly or be groomed off (fig. 7.7), or the fly may transport lice to a host on which they cannot establish a viable population. The answer to this apparent paradox is that phoresis may have adaptive trade-offs, as discussed earlier. It provides a way for lice to escape a dead or dying host. It also provides a way for wing lice to escape competition with body lice. In summary, despite the many similarities between wing and body lice, wing lice are much better at dispersing to new hosts than body lice.

8   POPULATION STRUCTURE

No self is an island; each exists in a fabric of relations
that is now more complex and mobile than ever before.
—Lyotard 1984

Populations are seldom uniform; they tend to be subdivided, with gene frequencies unevenly distributed across
landscapes. Population genetic structure arises largely
because of limitations to dispersal and gene flow. Population variation in parasites is influenced by the rate of
parasite dispersal at three scales: dispersal among host individuals, among
host populations, and among host species (fig. 8.1; Combes 2001; Huyse
et al. 2005). If most dispersal consists of vertical transmission from parent
hosts to their offspring, then parasites living on individual hosts and their
progeny will accumulate genetic differences, contributing to population
genetic structure. If horizontal transmission is common, however, it will
tend to erode population genetic structure among host individuals. It is
important to have information regarding the dispersal ecology of parasites
in order to interpret parasite population structure.

Differential selection and local adaptation among variable environments
can also play important roles in population structure, as do the effects of
small population size and inbreeding. It is important also to understand
these contributors to population structure in order to understand the
underlying causes of diversification and, ultimately, the origin of species.
Below we briefly review determinants of population structure in parasites.
We then focus on lice, which are excellent models for studies of population
structure.

Host individuals

Parasite ecologists use the term "infrapopulation" for conspecific parasites
living in, or on, a single host individual (Poulin 2007; Goater et al. 2013).
Parasites can show more or less structure among infrapopulations. If parasites disperse frequently among hosts, then the regular mixing of the parasite gene pool will prevent the emergence of much population genetic
structure among infrapopulations. In contrast, parasites that seldom move
between hosts will have infrapopulations with greater population genetic
structure.

Infrapopulations linked by periodic gene flow make up a metapopula-

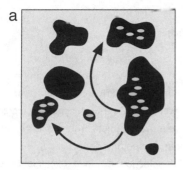

FIGURE 8.1. Comparison of (a) free-living metapopulation to (b) parasite metapopulation. In panel a, some habitat patches are occupied by individual organisms (grey dots), while other patches are unoccupied. Reproduction occurs within patches, with occasional dispersal between patches (arrows). In panel b, some bird hosts are occupied by parasites (grey dots), while other birds are parasite-free. Parasite reproduction occurs within the infrapopulations on individual birds, with occasional dispersal between infrapopulations (arrows).

tion—that is, a "population of populations" (fig. 8.1; Levins 1969). Metapopulation theory, developed mainly by ecologists (Hanski 1998; Krebs 2009), was not widely applied to host-parasite systems until relatively recently (Gandon 2002; Huyse et al. 2005). Metapopulations have several emergent properties that can dramatically influence the population genetic structure of parasites. For example, infrapopulations that are part of a metapopulation are less likely to go extinct because the probability of recolonization is much higher (Futuyma 2013). Additionally, parasites in different infrapopulations experience different types and magnitudes of selection and drift. Gene flow between infrapopulations within a metapopulation can provide an influx of genetic variation that fuels coevolutionary processes (Gandon et al. 2008).

An important aspect of metapopulations is that inbreeding within infrapopulations can contribute to genetic differentiation among infrapopulations of parasites on different host individuals (box 8.1). Small effective population sizes are likely to be common among parasite infrapopulations. Population genetic theory predicts that larger population sizes will show less evidence of inbreeding, and less population structure (Futuyma 2013). Nadler (1995) tested this prediction for parasites by comparing the population genetic structure of livestock nematodes (*Ostertagia ostertagi* and *Ascaris suum*) with similar life cycles. *O. ostertagi*, which has large infrapopulations (thousands of worms per host), showed less population structure than *A. suum*, which has small infrapopulations (<100 worms per host.) The results of this study are consistent with the prediction that inbreeding in small parasite infrapopulations leads to greater population structure.

BOX 8.1. Parasite population genetics

Inbreeding of parasites on hosts has important consequences for the population structure of neutral genetic variation. It also can have profound consequences for the genetic variation of non-neutral variation, particularly in relation to random genetic drift (Ohta 1992). Inbreeding can lower population-level fitness by increasing the probability that deleterious recessive mutations get combined in the same individual (Muller 1964). In such populations, inbreeding depression will be pronounced. To overcome inbreeding depression, selection may favor dispersal to new hosts. Alternatively, over long timescales, recessive deleterious mutations may be purged as a result of increased exposure to selection in small populations with more homozygous mutations.

Because deleterious mutations vastly outnumber beneficial mutations, theory suggests that deleterious mutations will accumulate more in small populations than large ones (Ohta 1992; Gillespie 2001). This effect is especially pronounced for haploid genes, such as those in mitochondria. Given the low effective population sizes of permanent parasites on single host individuals, random genetic drift can lead to increases in the accumulation of deleterious mutations. The substitution rate for such parasites may be higher than in free-living organisms because more deleterious mutations go to fixation in small populations, compared to large populations (Johnson and Seger 2001).

On the other hand, structured populations effectively increase effective population size (Wright 1943). It may be, therefore, that the structuring of populations into hundreds and thousands of small infrapopulations increases effective population size. However, a recent study of whale lice, which are amphipod (not insect) associates of whales with overall population structure similar to that of bird and mammal lice, found that effective population sizes inferred from mitochondrial data were much lower than expected (Seger et al. 2010). The result was explained by selection on forward and backward mutations. Dispersal is also important in determining the fate of mutations in the face of genetic drift. More theoretical models based on the population genetic structure of permanent parasites, such as lice, are needed to understand the effects of infrapopulation structure on genetic substitution rates.

The prediction that deleterious mutations will accumulate faster in small populations suggests that parasites will have higher substitution rates at the molecular level than their free-living counterparts. Several studies have shown that parasites do indeed have faster rates of nucleotide substitution than hosts, or free-living relatives (Downton and Austin 1995; Page et al. 1998; Young and dePamphilis

2005; Hassanin 2006; Lemaire et al. 2011). Faster rates of substitution are particularly evident for mitochondrial or chloroplast genes, which are haploid and do not recombine. Other explanations have been proposed, such as coevolutionary arms races (Aguileta et al. 2009), faster generation times (Hafner et al. 1994; Page et al. 1998), warmer thermal environments (Rand 1994; Simmons and Weller 2001), relaxed selectional constraints (Young and dePamphilis 2005; Lemaire et al. 2011), and increased mutation rates (Young and dePamphilis 2005). While these alternate explanations may be plausible in some cases, they do not have the generality that the population genetic theory of deleterious mutations provides, assuming parasites do indeed have smaller effective population sizes than their free-living relatives.

Host populations

If dispersal of parasites among infrapopulations is common, then the parasites across individual hosts in a host population will represent a single interbreeding population of parasites, rather than a metapopulation of parasites. If parasites on different host populations undergo periodic extinction and recolonization from other host populations, then the parasites on different host populations will represent a metapopulation (Poulin 2007). This scale of dispersal and population structure has not been well characterized because dispersal of parasites among host populations is even more difficult to monitor than dispersal of parasites between host individuals (Huyse et al. 2005). Dispersal of the hosts themselves may be the most common mechanism by which parasites disperse among host populations (Boulinier et al. 2001; McCoy et al. 2003; Criscione et al. 2005).

Despite the difficulties of tracking the movement of parasites among host populations directly, parasite dispersal among host populations can be inferred by comparing the phylogeographic structure of the parasite and the host (Criscione et al. 2005; Huyse et al. 2005). If host dispersal is the main way in which parasites disperse between host populations, then the population genetic structure of the parasites will mirror that of the hosts (Rannala and Michalakis 2003). A "cophylogeographic" study based on mitochondrial sequences of the nematode parasite *Heligmosomoides polygyrus*, and its field mouse host (*Apodemus sylvaticus*), provides an example of parasite and host phylogeographic topologies that are more similar than expected by chance (Nieberding et al. 2004). We consider examples involving lice and their hosts later in this chapter.

Host species

Relatively few parasites are specific to a single species of host; most parasites use at least two or more host species. For many parasites it is easier to disperse to another individual of the same host species than to an individual belonging to a different host species. This is particularly true of parasites in which vertical transmission is predominant. In such parasites there will certainly be more dispersal of parasites within host species than between host species, leading to the possible formation of host races of parasites. In the case of permanent parasites, such as lice, dispersal limitations appear to have a major influence on the evolution of host specificity and diversification, as we discuss below.

Selection can also limit the ability of parasites to survive and reproduce on different host species, owing to local adaptation (Gandon and Michalakis 2002). Host race formation in herbivorous insects has been studied mainly from this perspective (Filchak et al. 2000; Servedio et al. 2011). Selection and dispersal limitations can work together to govern host-specificity, and both processes may contribute to the formation of host races and, ultimately, to the speciation of parasites (Rundell and Price 2009; Servedio et al. 2011). The relative importance of dispersal limitations and selection in the diversification of parasites is a topic of widespread interest (Gandon and Michalakis 2002; Huyse et al. 2005; Thompson 2005; Schmid-Hempel 2011). Permanent parasites, such as lice, are powerful models for work on these topics because of the relative tractability of their populations across different scales.

Population structure of lice among host individuals

Because lice pass multiple generations on a single host individual, there can be significant genetic structure among the lice on different hosts. In some cases, such as the lice of pigeons and doves (Bush and Clayton 2006), effective population sizes are small enough (<25 lice per bird) to also make random genetic drift a potentially important factor. However, periodic dispersal of lice between hosts will mitigate the effects of drift. Thus, if there is any population structure of lice among individual hosts, it is presumably due mainly to dispersal limitations.

Most knowledge of the population structure of lice among host individuals is from a large body of work on the chewing lice of North American pocket gophers (Hafner et al. 2003). Gophers are burrowing mammals that spend much of the their time underground, with individual gophers seldom coming into contact. Nadler et al.'s (1990) study of allozyme differentiation in *Geomydoecus actuosi* gopher lice documented significant deficiency of heterozygotes in 4 of 13 infrapopulations measured. Infrapopulations of gopher lice within the same geographic locality revealed significant differentiation (fixation index [F_{ST}] values 0.039 to 0.162), consistent with extensive inbreeding on single host individuals.

Inbreeding has also been inferred for human head and body lice, despite the fact that humans come into more regular contact than gophers. Leo et al. (2005) studied lice from humans infested with body lice and head lice, including a pair of sisters with shared sleeping quarters. A survey of microsatellite markers revealed genetic differentiation between the head louse populations, and between the body louse populations, on the two sisters. An analysis of migration using genetic data from 14 individual lice suggested that only one head louse and three body lice had migrated between the sisters. Thus, even lice on host individuals that come into regular contact can show evidence of inbreeding on individual hosts.

A recent study of feather lice (*Degeeriella regalis*) on Galápagos hawks (*Buteo galapagoensis*) demonstrated significant population genetic structure among the lice on individual birds (plate 15a,b; Koop et al. 2014). Louse infrapopulations were genetically differentiated in 93% of pairwise F_{ST} comparisons among individual hawks on the island of Fernandina, and 67% of comparisons on the island of Santiago. Thus, individual hawks essentially represent islands with limited gene flow among the infrapopulations of *D. regalis* living on them.

Population genetic data also have the power to shed light on modes of transmission within single host populations. Dispersal of pocket gopher lice was once assumed to occur largely, if not solely, by vertical transmission between mother hosts and their offspring. This idea was supported by the observation that populations of lice on female pocket gophers decline by more than 50% when females give birth to their offspring, presumably owing to loss of lice by vertical transmission (Rust 1974). However, population genetic studies of lice in a gopher species hybrid zone (Patton et al. 1984; Demastes et al. 1998) have revealed that lice associated with gopher individuals do not, in fact, sort according to host mitochondrial haplotype lineages (a maternally transmitted genetic marker). Therefore, the assumption of strict vertical transmission of gopher lice is apparently not correct. Some horizontal transmission of the lice must also be occurring.

Population structure of lice among host populations

Lice are expected to show genetic structure among host populations since, by definition, host populations and their permanent parasites are spatially isolated. In support of this prediction, allozyme studies of pocket gopher lice showed extensive genetic differentiation among populations of pocket gophers (*Thomomys bottae*) and their lice (*Geomydoecus actuosi*) separated by 20–300 km (Nadler et al. 1990). The extent of genetic differentiation was very similar for the gophers and lice (F_{ST} = 0.236 for gophers and F_{ST} = 0.240 for lice). These results show that, at a landscape scale, parasite gene flow can track host gene flow.

Deep population genetic breaks in hosts are sometimes also observed in their lice. For example, head lice (*Philopterus* sp.) on Australian magpies (*Gymnorhina tibicen*) show an east-west genetic break that corresponds to an east-west genetic break in the birds themselves (Toon and Hughes 2008; Whiteman 2008) (fig. 8.2). However, there is also a clade of *Philopterus* sp. in central Australia, despite the fact that the magpies have no such clade (Toon and Hughes 2008). Moreover, in contrast to the *Philopterus* results, there is a deep north-south population genetic break in the louse *Brueelia semiannulata* from the same birds. These results show that gene flow in lice does not necessarily reflect gene flow in hosts (Toon and Hughes 2008; Whiteman 2008). A possible explanation for the different patterns may be that the lice are not strictly host specific. They may occur on other hosts that also influence their gene flow.

Another study of the cophylogeographic structure of lice across host populations involves the Galápagos hawks mentioned earlier. In addition to *Degeeriella regalis*, the hawks have two other species of lice: *Colpocephalum turbinatum* and *Craspedorrhynchus* sp. (Whiteman et al. 2007, 2009). There is little mitochondrial DNA variation, and therefore no mtDNA genetic structure, among the populations of hawks on different islands (plate 15b,c). In contrast, there is considerable variation in the mitochondrial DNA of all three species of lice among the hawk populations. The ischnoceran louse *D. regalis* shows the most variation, with the variation structured among the islands. F_{ST} values for *D. regalis* are positively correlated with estimates of gene flow for the hawks based on more variable microsatellite data (Whiteman et al. 2006). Populations of *D. regalis* also show pronounced genetic variation among individual hawks, as discussed in the previous section of this chapter.

Mitochondrial gene sequences for the ischnoceran louse *Craspedorrhynchus* sp. (Whiteman et al. 2009), and the amblyceran louse *Colpocephalum tubinatum* (Whiteman et al. 2007), also show significant population genetic structure among populations of hawks on different islands. The population genetic structure is less in these species of lice than in *D. regalis*, and it is not as well correlated with host population genetic structure as *D. regalis*. Since (non-phoretic) ischnoceran lice are less mobile than amblyceran lice, Whiteman et al. (2009) predicted the cophylogeographic structure between hawks and the ischnoceran louse *Degeeriella* would be more congruent than in the case of the amblyceran louse *Colpocephalum*. However, differences in

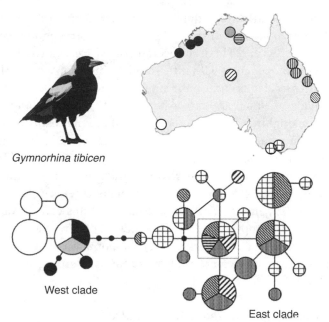

Gymnorhina tibicen

West clade

East clade

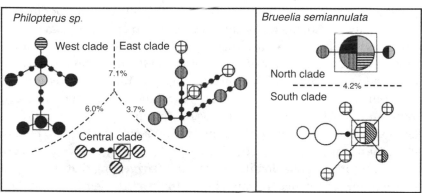

Philopterus sp.

West clade | East clade

7.1%

6.0%   3.7%

Central clade

Brueelia semiannulata

North clade

- - - - - - - - 4.2% - - - - - -

South clade

FIGURE 8.2. Haplotype networks based on mtDNA from Australian magpies (*Gymnorhina tibicen*) and their two species of chewing lice: *Philopterus* sp. and *Brueelia semiannulata*. The sampling locality for each haplotype corresponds to the pattern on the map. The size of each circle represents the number of individual birds or lice sampled (small n < 5, medium n = 5–20, large n > 20). Inferred ancestral haplotypes, based on their frequencies and position in the network, are indicated by boxes. After Toon and Hughes (2008).

mobility cannot explain why the other ischnoceran louse *Craspedorrhynchus* sp. also showed less divergence and structure than *Degeeriella*. This question requires further investigation.

Štefka et al. (2011) performed a similar comparison of the genetic differentiation of two lineages of lice on mockingbirds (*Mimus* spp.) across the

Galápagos archipelago. However, only the genetic structure of the amblyceran louse (*Myrsidea nesomimi*) showed significant correspondence with host population structure. The ischnoceran louse (*Brueelia galapagensis*) was genetically homogeneous, with little standing variation across islands. *Brueelia galapagensis* also occurs on the small ground finch (*Geospiza fuliginosa*), individuals of which are known to move among the islands more readily than mockingbirds. This difference in host use may explain the lack of genetic structure in *B. galapagensis* among populations of mockingbirds across different islands (Sari et al. 2013).

The examples above illustrate cophylogeographic studies of lice and their hosts. Because the DNA substitution rates of parasites are often higher than those of hosts, genetic information from parasites can be more informative than that from hosts over a given period of time. In cases such as this, lice can be used as markers to infer the population structure and dynamics of the hosts (Whiteman and Parker 2005; Nieberding and Olivieri 2007).

Population structure of lice among host species

Lice that use more than one host species are likely to show population genetic structure among the host species, unless dispersal is common, in which case the lice may reflect the same barriers to gene flow that influence the hosts. An example of this is provided by the ischnoceran louse *Austrophilopterus cancellosus.* This louse occurs on two to three species of toucans at each of several different geographic regions in Central and South America (fig. 8.3; Weckstein 2004). Specimens collected from different species or subspecies of toucans within a given region often show little genetic variation. Interestingly, however, the lice on opposite sides of major Amazonian rivers do differ genetically, suggesting that rivers limit the gene flow of the lice, presumably by limiting the gene flow of their hosts (fig. 8.3). Rivers are indeed barriers for toucans and other birds; different subspecies of a given toucan species often occur on the opposite sides of major rivers.

The population genetic structure of parasites that exploit more than one host species can also be influenced by differential selection, or some combination of selection and dispersal. Experimental transfers of pigeon lice to novel host species show that lice can survive and reproduce on novel hosts only if they are not too different in size from the native host (chapter 5; Bush and Clayton 2006). Thus, among different-sized (sympatric) hosts, dispersal limitations will have less influence than selection on population genetic structure. Hence, when exploring patterns of population structure in lice, both selection and dispersal should be taken into account.

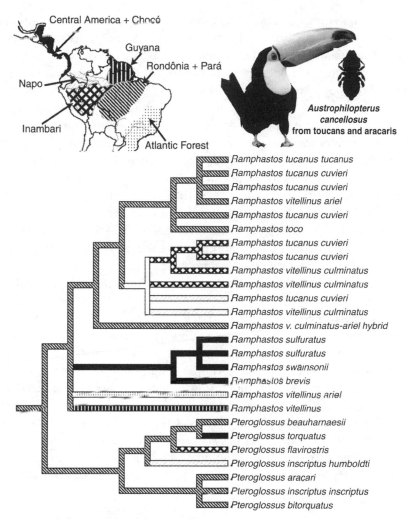

FIGURE 8.3. Phylogeography of the toucan louse *Austrophilopterus cancellosus*, based on mitochondrial (COI) and nuclear (EF1-alpha) genes. The geographic regions mapped onto the tree reflect those shown on the map. Each branch is labeled with the host species or subspecies from which the specimen of *A. cancellosus* was collected. Note that the lice are more clustered by region than they are by host species or subspecies. After Weckstein (2004).

Population structure of ecological replicate lice

Ecological replicates that vary in dispersal ability, such as pigeon body lice and wing lice, provide an opportunity to explore the influence of dispersal on the population genetic structure of related groups (Johnson and Clayton 2003b). Body lice would be predicted to have more population genetic structure than wing lice, assuming phoresis plays a major role in wing louse dispersal. Since body lice are not phoretic, their populations should show more structure. Differences in population structure can occur among host individuals, host populations, or host species. Unfortunately, no careful study of the population structure of body and wing lice among host individuals has been carried out. Because several species of pigeons and doves with host-specific lice are common, large numbers of hosts can be sampled, providing infrapopulations of both wing and body lice to compare among host individuals, populations, and species. A study of this kind is needed.

The relative contributions of horizontal and vertical transmission to population structure can be quantified by regressing the degree of genetic relatedness of lice on the genetic relatedness of their hosts. If these parameters are highly correlated, then it suggests that vertical transmission is the major mode of dispersal. Depending on the degree of inbreeding, population genetic analyses might even detect individual lice that are migrants, as in the case of the human head and body lice from sisters discussed above (Leo et al. 2005).

The population genetic structure of wing and body lice has been compared across different geographic regions. In a study comparing mitochondrial gene sequences of lice around the Gulf of Mexico (Johnson et al. 2002b), body lice (*Physconelloides*) showed more genetic differentiation than wing lice (*Columbicola*) over the same geographic distance (fig. 8.4). For example, haplotypes of *Physconelloides ceratoceps* were never shared between two widely separated localities in Texas, USA, and Campeche, Mexico. In contrast, two of the haplotypes of *Columbicola macrourae* occurred on doves in both Texas and Campeche. To our knowledge, no other study comparing wing and body lice across geographically separated populations has been conducted.

Studies of population structure in wing versus body lice have focused mainly on populations of lice that live on more than one species of host. This scale of comparison is important because it relates to host race formation and the evolution of host specificity. Based on conventional taxonomy using morphological traits, neither wing lice (*Columbicola* spp.) nor body lice (*Physconelloides* spp.) are absolutely host specific; several species in each genus are found on multiple host species (Price et al. 1999; Clayton and Price 1999). However, as expected from differences in their dispersal ecology, body lice are still significantly more host specific than wing lice (Johnson et al. 2002b).

Phylogenetic reconstructions using mitochondrial sequences of wing lice (*Columbicola*) and body lice (*Physconelloides*) reveal the presence of cryptic species (Johnson et al. 2002b). Genetic divergence between populations of

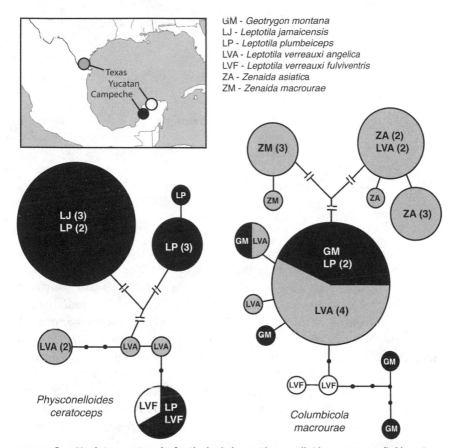

GM - *Geotrygon montana*
LJ - *Leptotila jamaicensis*
LP - *Leptotila plumbeiceps*
LVA - *Leptotila verreauxi angelica*
LVF - *Leptotila verreauxi fulviventris*
ZA - *Zenaida asiatica*
ZM - *Zenaida macrourae*

Physconelloides ceratoceps

Columbicola macrourae

FIGURE 8.4. Haplotype networks for the body louse *Physconelloides ceratoceps* (left), and the wing louse *Columbicola macrourae* (right) sampled from six dove species around the Gulf of Mexico. Networks are based on mitochondrial COI sequences. Each connection represents a single mutational step, with small bullets representing inferred (unsampled) haplotypes. Observed haplotypes are shown as circles, with circle size indicating the relative number of individuals sampled with that haplotype. Shading corresponds to the collecting localities shown on the map, and indicates the origin of each sampled louse. Host associations are indicated with the abbreviations defined in the legend. When more than one louse was sampled from a particular haplotype, the number of lice is indicated in parentheses. Deep divergences (indicated by double hashes) are potentially cryptic species of lice. After Johnson et al. (2002b).

lice on different host species is extremely high in some cases. In general, cryptic species are more common in body lice than wing lice, although there are also some examples of cryptic species in wing lice (Malenke et al. 2009). One particularly widespread species, *Columbicola macrourae*, shows a mixed pattern. Several highly divergent mitochondrial haplotypes have

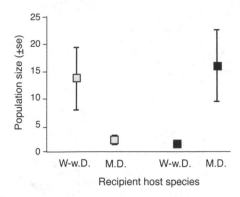

FIGURE 8.5. Experiment in which *Columbicola macrourae* haplotypes from white-winged doves (white squares) and mourning doves (black squares) were transferred between the two hosts (and to control hosts). Points show population sizes at the conclusion of the experiment. Each louse had significantly higher fitness on its native host, compared to the novel host (Wilcoxon signed rank test, P < 0.05) (from Malenke et al. 2009). The experiment ran for two months.

been detected within this species. Some of these are restricted to a single host, while others are distributed across three host species (fig. 8.4).

Columbicola macrourae populations on mourning doves and sympatric white-winged doves are genetically distinct from one another, and distinct from *C. macrourae* on other hosts. These genetic differences appear to be a result of host race formation due to local adaptation (Malenke et al. 2009). White-winged doves (mean 153 grams) are 10% larger than mourning doves (mean 139 grams) and their lice are larger than mourning dove lice. These differences are ecologically meaningful, because reciprocal transfer experiments between the two hosts showed a decrease in the survival and reproductive success of lice on the "wrong" host (fig. 8.5; Malenke et al. 2009). White-winged doves and mourning doves are sympatric and have the same species of hippoboscid fly (Maa 1969). Thus, it is likely that cryptic speciation in these populations has been driven by host race formation and selection for size differentiation on the different species of doves, not simply by dispersal limitations.

Putting it all together: The geographic mosaic

Acting together, selection and dispersal are responsible for much of the mosaic pattern of coevolution across geographic space, as described in the geographic mosaic theory of coevolution (fig.1.5; Thompson 2005). Selection may differ between different geographic locations, creating a selection mosaic. Dispersal and gene flow will dictate much of the amount of "trait-remixing." These processes may result in coevolutionary hotspots, where reciprocal selection is strong, and coldspots, where reciprocal selection is weak.

Several features of the bird-louse interaction could promote a geographic mosaic of coevolution. For example, one potential mosaic in the selection regime is geographic variation in humidity. Feather lice are intolerant of low humidity (fig. 2.5; Moyer et al. 2002a). Indeed, in regions of very low humidity, lice may be completely absent, creating a coevolutionary coldspot in which selection for host defense will be reduced. As humidity increases, louse prevalence increases. Humidity can also influence the outcome of competition in lice (chapter 5, and Malenke et al. 2011). Thus coevolutionary outcomes resulting from competition are also expected to vary across geographic humidity gradients.

Another possible contributor to a geographic mosaic concerns localized selection on non-specific lice occurring on different host species. Because the range distributions of different host species are almost never perfectly overlapping, there will be regions of the parasite's geographic distribution in which one or more of its hosts is absent. For hosts that differ slightly in body size, selection should favor a closer match in geographic regions where only one host is present. For example, Inca doves (*Columbina inca*) and common ground doves (*Columbina passerina*) share the same species of wing lice (*Columbicola passerinae*) (Clayton and Price 1999). Inca doves are slightly larger than common ground doves, and in regions where both hosts occur, selection may be different for parasite populations on different host species. Thus, there may be a selection mosaic, with host-mediated selection differing depending on which host species a louse is on. Dispersal of parasites between hosts may lead to gene flow that, depending on its magnitude, could erode local adaptation to particular host species. Alternatively, there may be some local adaptation, but over time gene flow may be sufficient to prevent host-race formation and speciation. An interesting study would be to compare the body sizes of *Columbicola passerinae* on Inca doves and common ground doves in regions where both hosts are present, and in regions where one or the other host is absent.

The very nature of mutation, genetic drift, selection, and dispersal will also lead to a geographic mosaic in the pattern of genetic variation for both hosts and parasites. In some ways, this mosaic patterning might be unpredictable because of the stochastic nature of mutation, genetic drift, and dispersal. For pigeon wing lice, one approach to control for this stochasticity would be to compare the genetic structure of lice in pigeon populations where hippoboscid flies are present, versus populations where flies are absent. In addition, because pigeon lice can breed for many generations on a single host individual, there may even be opportunities for local adaptation to single host individuals. Differences in dispersal and gene flow for lice in pigeon populations with and without hippoboscid flies could mediate the amount of local adaptation. Variation in dispersal opportunities for lice may allow more general patterns to be detected.

Comparisons of patterns across populations of wing lice to those across populations of body lice on the same hosts could be informative. If similar patterns are observed in wing lice and body lice in populations *without* flies, compared to different patterns in populations *with* flies, then one might conclude even more firmly that dispersal via phoresis influences geographic mosaic patterns of gene flow and selection in wing lice.

In summary, local adaptation and dispersal limitations can contribute to population genetic structure that leads to more substantial diversification and speciation of parasites. We turn our attention to these two topics in the final chapters of this book.

IV CODIVERSIFICATION

9   COPHYLOGENETIC DYNAMICS

Evolution is an obstacle course not a freeway.
—Gould 1993

One of the least understood aspects of evolutionary biology is how microevolutionary processes translate into broad patterns of lineage diversification (Mayr 1963; Sobel et al. 2010; Thompson 2013). How do mutation, selection, dispersal, and random genetic drift create genotypic and phenotypic diversity over time? Moreover, what are the ecological factors that influence these processes? Diversification can be favored by selection, with new lineages occupying new adaptive zones, or it can be nonadaptive (Schluter 2009; Rundell and Price 2009). For speciation to occur there must be limited gene flow between diverging populations. Barriers to dispersal ultimately result in barriers to reproduction, which cause lineages to diverge (Sobel et al. 2010). Lineage diversification may also be enhanced because of selective differences across the barrier (Rundell and Price 2009). Yet even in the absence of selection, isolated populations will ultimately diverge through the ongoing action of mutation and genetic drift (Orr 1995). With or without differences in selection, barriers to reproduction are key to understanding why populations diversify.

Interacting groups with patterns of codiversification are powerful arenas for testing the influence of selection, dispersal, and other processes on lineage diversification. When reproduction in one group is linked to reproduction in another group, codiversification may occur (box 1.1). For example, the movement of pollinators between individual plants links those plants into a single interbreeding population. If a barrier to pollinator movement arises, gene flow between the plants will be disrupted. As another example, parasite dispersal is often linked to host dispersal, with barriers to host movement also influencing the movement of their parasites. This linkage of host and parasite dispersal is particularly common in permanent parasites, such as lice, which complete all stages of their life cycle on the body of the host. If barriers to movement contribute to lineage diversification in both host and parasite, then they will codiversify (box 1.1). If codiversification is simultaneous, then the host and parasite may undergo cospeciation.

Repeated bouts of cospeciation can generate congruent phylogenies (fig. 9.1). Cophylogenetic approaches are used to assess the amount of con-

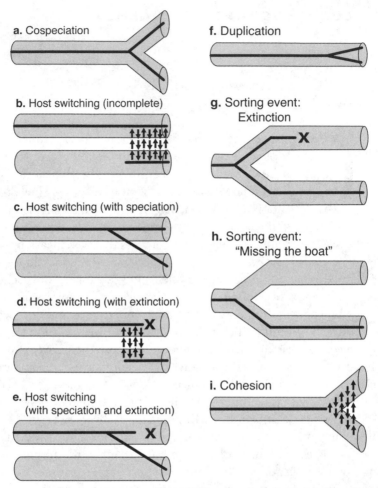

a. Cospeciation

f. Duplication

b. Host switching (incomplete)

g. Sorting event:
Extinction

c. Host switching (with speciation)

h. Sorting event:
"Missing the boat"

d. Host switching (with extinction)

e. Host switching
(with speciation and extinction)

i. Cohesion

FIGURE 9.1. Macroevolutionary events that influence the phylogenetic interface. Pipes represent hosts and black lines represent parasites. Arrows represent gene flow and X's represent extinction. (a) Cospeciation, in which the host and parasite speciate at the same time, yielding congruent phylogenies; (b) Incomplete host switching, in which a parasite colonizes a novel host, but maintains gene flow with parasites on the original host; (c) Host switching with speciation; (d) Host switching in which the parasite colonizes a new host, then goes extinct on the original host; (e) Host switching in which the parasite colonizes a new host, speciates, then goes extinct on the original host; (f) Duplication of parasites on a single host; (g) Cospeciation followed by extinction of one parasite; (h) "Missing the boat," in which the parasite fails to colonize one of two diverging host lineages; (i) Parasite cohesion, in which a parasite maintains gene flow between diverging host populations.

BOX 9.1. Preferential host switching and pseudocospeciation

Like cospeciation, some forms of host switching can reinforce congruence between phylogenies (Brooks 1979; de Vienne et al. 2007). If host resources are correlated with host phylogeny, parasites are more likely to switch to related hosts than unrelated ones. For example, if plant chemistry is conserved within plant clades, but differs among clades, then herbivorous insects are more likely to switch between plant species within the same clade. Computer simulations show that preferential host switching can generate congruent phylogenies and thus pseudocospeciation. However, in the simulations congruence occurs most often when the host phylogeny is balanced (symmetrical).

The risk of being misled by pseudocospeciation means that it is essential to distinguish true cospeciation from preferential host switching. One way to do this is to evaluate the relative timing of host and parasite diversification (Charleston and Robertson 2002; de Vienne et al. 2007). Cospeciation is simultaneous diversification. In contrast, preferential host switching causes parasite diversification that lags behind host diversification. The relative timing of diversification can be tested using phylogenies that are time calibrated using molecular dating techniques. This approach has been applied to a variety of systems, such as herbivorous insects of plants (Percy et al. 2004), mildews of maple trees (Hirose et al. 2005), and viruses of primates (Charleston and Robertson 2002; Cuthill and Charleston 2013). In all of these cases, parasite radiation follows host radiation, implicating preferential host switching.

gruence, and to test for macroevolutionary events, such as cospeciation or host switching, that influence congruence. Some authors use "cophylogeny" as a synonym for congruence (e.g., Hafner et al. 2003; Althoff et al. 2014). However, like most authors, we prefer to reserve the term "cophylogenetics" for the practice of making cophylogenetic comparisons, regardless of the extent of congruence (de Vienne et al. 2013). As discussed in chapter 1, congruent phylogenies do not necessarily imply cospeciation. They may be due to pseudocospeciation (chapter 1). For example, congruent phylogenies can be generated by preferential host switching (box 9.1; de Vienne et al. 2007), also known as clade-limited colonization (Sorenson et al. 2004).

Even if congruent phylogenies are caused by cospeciation, this does not provide evidence of reciprocal selection or coadaptation (chapter 1). Coadaptation does not require cospeciation, nor does cospeciation require coadaptation (Janz 2011). In fact, some processes involving coadaptation, such as escape-and-radiate coevolution, involve a rapid burst of diversification in one group, followed by a reciprocal burst of diversification in a second group. This is very different than cospeciation and will not lead to congruent phylogenies (Segraves 2010; Thompson 2013). The utility of congruent phylogenies, when they occur, is that they confirm prolonged associations between interacting groups over macroevolutionary time. Prolonged associations can be useful frameworks for studies of coadaptation and the influence of coadaptation on codiversification (see chapters 10 and 11).

The goal of host-parasite cophylogenetic comparisons is to understand which processes make phylogenies more congruent, and which processes make them less congruent. De Vienne et al. (2013) provided a thorough review of analytical methods for comparing host and parasite trees, and they assessed the relative importance of different events in the history of host-parasite associations. Five main processes govern cophylogenetic history: cospeciation, host switching, duplication, sorting events, and what we term cohesion (fig. 9.1). We consider each of these processes in detail below. We then describe a case study in which the processes have interacted to mold patterns of codiversification between primates and their lice. Lice and their hosts are quite literally textbook examples of cophylogenetic dynamics (Bergstrom and Dugatkin 2011; Futuyma 2013; Zimmer and Emlen 2013; Herron and Freeman 2014).

Cospeciation

Cospeciation is the synchronous speciation of ecologically interacting groups (box 1.1; fig. 9.1a). Repeated bouts of cospeciation generate congruent phylogenies. For example, the phylogeny of the body louse genus *Physconelloides* matches that of mid-sized New World doves in the genus *Zenaida* (fig. 9.2). Congruent phylogenies link interacting lineages over macroevolutionary time. This linkage provides a useful framework for tests in both micro- and macroevolutionary time. For example, congruent phylogenies can be used to make comparisons of rates of molecular evolution between hosts and parasites, as we show later in this chapter.

Although investigators and students often find the process of cospeciation to be fascinating, it is far from the dominant process in host-parasite codiversification. Indeed, highly congruent phylogenies are relatively uncommon (Brooks and McLennan 1991; Tellier et al. 2010; de Vienne et al. 2013). Despite this fact, congruent phylogenies have been documented in an impressive range of host-parasite systems, including mammal viruses (Jackson and Charleston 2004; Switzer et al. 2005), mammal pinworms (Hugot 1999, 2003), bat mites (Bruyndonckx et al. 2009), avian feather mites (Dabert et al. 2001; Dabert 2003), marsupial nematodes (Chilton et al. 2011), cope-

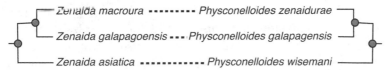

FIGURE 9.2. Cophylogenetic reconstruction of body lice (right) with dove hosts (left). In this example the lice have cospeciated with the hosts twice, yielding congruent phylogenies. Bullets are inferred cospeciation events. After Johnson and Clayton (2004).

pods of fish (Paterson and Poulin 1999), microsporidia of mosquitoes (Baker et al. 1998), and gall-forming nematodes of grasses (Subbotin et al. 2004).

Certain groups of lice show congruent phylogenies with their hosts, ranging from the chewing lice of marine birds (Page et al. 2004; Hughes et al. 2007) to the sucking lice of heteromyid rodents (Light and Hafner 2008). However, many groups of lice show no congruence, including lice from rock wallabys (Barker 1991, 1994), toucans (Weckstein 2004), penguins (Banks et al. 2006), and Peruvian rodents (V. S. Smith et al. 2008). This variation in cophylogenetic history within a single order of insects is because of different mixtures of cospeciation, host switching, duplication, sorting events, and cohesion (fig. 9.1). Understanding why some groups have congruent phylogenies, while others do not, can provide insights into the role of different ecological processes that influence patterns of codiversification.

In host-parasite systems, the parasites are often thought of as responding to the host, given that their survival and reproduction are dependent on the host, while the host can survive perfectly well, if not better, without the parasite. One reason why fully congruent phylogenies are rare, even in host-parasite systems, is because repeated bouts of cospeciation require host-specific parasites that are restricted to one species of host (Humphrey-Smith 1989; Page 2006). Most parasites show a broader range of host use, from moderate specialists to generalists. It is important to realize that host specificity is not equivalent to cospeciation or congruence; host specificity is a necessary, but insufficient, condition for cophylogenetic congruence (Hoberg et al. 1997; Clayton et al. 2004). Most members of the louse genus *Brueelia* are very host specific, yet they show no evidence of cospeciation with their hosts (see chapter 10).

Cospeciation is most likely when parasite reproduction is tightly linked to host reproduction (Thompson 2005). This linkage can occur in several ways. If transmission of parasites occurs primarily during host reproduction, then factors isolating host individuals and populations reproductively may also isolate their parasite populations. For example, sexually transmitted parasites rely on male and female contact during copulation, and thus the dispersal of parasites and their resulting population structure will be directly tied to reproductive contacts between host individuals (Jackson 2005). For other parasites, the primary mode of transmission is vertical, from parent to offspring. Vertical transmission also provides a link between host and parasite reproduction (Page 2006). When vertical transmission is predominant, parasite genealogies will be a reflection of host genealogies. Thus, factors that isolate host populations will de facto isolate their parasites, setting the stage for cospeciation.

Even if parasite reproduction is not directly tied to host reproduction, the population structure of hosts and parasites can still be linked, as discussed in the last chapter. If parasite dispersal is closely tied to host dispersal (McCoy et al. 2003), then parasite population structure will mirror host population structure, favoring cospeciation. This is more common in the case of permanent parasites, such as lice, which spend their entire life cycle on the host.

The previous examples show how cospeciation can be governed by correlated dispersal and population structure of parasites and hosts. However, cospeciation can also be reinforced by selection, even when parasites are capable of dispersing independently of their hosts. If local adaptation to differences among host populations is common, parasite populations on different host populations can diverge even if there is no barrier to dispersal between host populations.

For example, many herbivorous insects fly between host populations. If these host populations differ in their defensive compounds against the insects, this can select for local adaptation by the insects to local plant chemistry (Laukkanen et al. 2012). Gene flow may then be restricted between diverging herbivore populations, not because of dispersal limitations, but because of the reduced fitness of insects dispersing to host populations on which they are not well adapted (Garrido et al. 2012; Coley and Kursar 2014). The end result is that, when plant populations speciate, their herbivores speciate with them.

When cospeciation takes place, host-specificity remains the same. Thus, if a host-specific parasite cospeciates with the host, then its daughter lineages are also likely to be host-specific parasites of the newly diverged host lineages. In summary, cospeciation leads to congruence between host and parasite phylogenies, and it increases parasite diversity. However, cospeciation is not as pervasive as host switching in shaping the cophylogenetic histories of most groups (de Vienne et al. 2013).

Host switching
Host switching occurs when a parasite lineage begins to use a new host. The concept of host switching can be confusing because most host "switches"

FIGURE 9.3. Cophylogenetic reconstruction of body lice (right) with their small New World ground dove hosts (left). *Physconelloides eurysema* 2 is inferred to have switched to *Columbina passerina* with continued gene flow (an incomplete host switch). Bullets are inferred cospeciation events. After Johnson and Clayton (2004).

are actually host expansions—that is, cases in which a parasite colonizes a new host, while continuing to use the original host. Host switching takes several forms, depending on the nature of gene flow between parasite populations on the original host and the novel host (fig. 9.1b-e). In cases where host switching leads to speciation, it has the potential to generate considerable new diversity (Ehrlich and Raven 1964; Janz 2011).

The first step in a host switch is dispersal of parasites from the original host to a novel host. If the dispersing parasites are able to establish a viable breeding population on the novel host, then it will be a successful host switch. As mentioned above, when gene flow is maintained between populations of parasites on both the novel and original hosts, the switch is more of a range expansion, and the specificity of the parasite has effectively decreased (fig. 9.1b). This process is also known as "host switching without speciation" or "incomplete host switching" (Johnson and Clayton 2004). Host switching without speciation does not lead to an increase in parasite diversity. An example can be seen in the phylogeny of small New World ground doves and their body lice (*Physconelloides*), in which one species of louse switched hosts and now occurs on two host species (fig. 9.3).

In some cases, gene flow between parasite populations on the original and novel hosts is reduced. This can occur because of a reduced opportunity for dispersal of parasites between the two host species, or because of selection for local adaptation by parasites on the different host species. In such cases, the host-range expansion may be followed by speciation of the parasite populations on the two different host species (fig. 9.1c). This process is called "host switching with speciation" (Johnson and Clayton 2004), or "partial host switching" (Ronquist 2003). It leads to an increase in parasite diversity.

It is also possible that a host switch is later followed by extinction of the parasite population on the original host species. When this happens, the host switch is considered "complete" (Ronquist 2003)—that is, the parasite lineage has completely switched from one host lineage to another. Extinction of the parasite population on the original host can occur regardless of whether the host switch has resulted in speciation. These different events are referred to as "host switching with extinction" (fig. 9.1d) or "host switching with speciation and extinction" (fig. 9.1e) (Johnson and Clayton 2004). In both cases, parasite diversity and host-specificity remain the same. These events are difficult to distinguish after the fact because the parasite no longer exists on the original host. In theory, with a very detailed fossil record of host-parasite associations, it might be possible to distinguish them.

Alternatively, a detailed study of population genetics could conceivably provide details of past host switching events. If the host switch itself is a trigger for speciation, then there may be a population bottleneck in the parasite that switched hosts. Such a bottleneck could leave a genetic signature of reduced allelic diversity in the switching population. On the other hand, if the host switch is not immediately followed by parasite population isolation, but there is continued gene flow with the population on the original host, then no signature of a population bottleneck is expected. In practice, these scenarios are very difficult to distinguish because population bottlenecks can also occur for other reasons.

What are the ecological factors that favor host switching? The first step, dispersal to a novel host, is more likely, of course, when parasites do not depend on host dispersal for their own dispersal. Host-independent dispersal mechanisms include parasites with free-living stages, parasites with vectors, and parasites capable of dispersing by phoresis. Even for parasites capable of independent dispersal, switching requires hosts that are sympatric and syntopic (sharing habitat). Host switching is more likely between hosts that live in large populations with broad geographic ranges than it is between small populations of hosts with partially overlapping, restricted geographic ranges.

Establishment, the second step in host switching, is more likely if the novel host is poorly defended against the novel parasite. If the host has no prior history of exposure to the parasite, it may be poorly defended, particularly if specialized defenses are required to combat the parasite (Schmid-Hempel 2011). The lack of a specialized host defense can generate a coevolutionary coldspot (Thompson 2005). Generalized defenses may be better at blocking host switches. Establishment requires that the novel host represent a resource base that can be used by the parasite. Necessary resources include food, shelter, and other basic needs. Establishment is also favored by freedom from competition with resident species of parasites (chapter 7) (Hudault et al. 2001; Dillon et al. 2005). In summary, any consideration of the likelihood of host switching must take into account *both* dispersal and establishment as factors determining whether a given host switch is likely to take place.

Duplication

Duplication refers to speciation of a parasite in the absence of host speciation (fig. 9.1f). The process is analogous to gene duplication within a single species tree (Page and Charleston 1998). In combination with sorting events (see below), duplication can lead to incongruence between host and parasite trees in the same way that individual gene trees can be incongruent with a species tree. Duplication increases parasite diversity, yet the host-specificity of each parasite lineage remains unchanged. An example of parasite duplication occurs in the body lice of band-tailed pigeons (*Patagioenas fasciata*), where sister lineages occur on the same host species (fig. 9.4). Populations of band-tailed pigeons in North and South America show no genetic differentiation; however, their body lice exhibit pronounced genetic divergence, indicative of an incipient duplication event.

Duplication may seem unlikely, given its apparent similarity to sympatric speciation; however, there are several possible mechanisms that can lead to parasite duplication. First, if parasite populations are patchily distributed across the range of a host species, there may be breaks in parasite gene flow, even with no pronounced breaks in host gene flow. Gaps in parasite distribution may be coevolutionary coldspots that contribute to parasite population divergence and speciation on host populations that are allopatric, at least from the perspective of the parasite. Such a speciation event results in two parasite species sharing a single host species, even though the two parasite species are allopatric (Paterson and Poulin 1999). This may be the case for the body lice of band-tailed pigeons described above (fig. 9.4).

Parasite duplication can also occur in sympatry. While sympatric speciation of parasites on the same host may seem unlikely, many large-bodied hosts provide different microhabitats into which parasites can diversify. For example, comparisons of the phylogeny of monogenean gill parasites to the phylogeny of their fish hosts indicates that the parasites have diversified

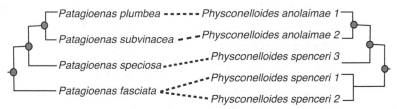

FIGURE 9.4. Cophylogenetic reconstruction of body lice (right) with their New World pigeon (*Patagioenas*) hosts (left). The body louse (*Physconelloides spenceri*) of band-tailed pigeons (*Patagioenas fasciata*) has diverged into two genetically highly distinct populations, potentially representing cryptic species, in which a duplication event is inferred. Bullets are inferred cospeciation events. After Johnson and Clayton (2004).

mainly through a process of parasite duplication on single host species (fig. 9.5; Simková et al. 2004). Using a phylogeny with 51 taxa, Guilhem et al. (2012) documented 11 cases in which sister species of parasites partition microhabitats within the gill structures of a single species of host. The results suggest that competition between parasites has driven ecological speciation on single host species, providing an example of sympatric duplication.

Parasites can also diverge in the timing of the life history stage at which they use the host. An example of this occurs in parasitic wasps of figs (Weiblen and Bush 2002). Parasitic wasps deposit their eggs inside figs by piercing the outside of the fig with a long ovipositor and inducing gall formation in the fig's ovaries, in which the larvae then develop. Comparison of the phylogenies of parasitic wasps and fig hosts reveals three cases in which sister species of wasps share the same species of fig, suggesting duplication events. In each case, the sister species differ in the length of their ovipositors (fig. 9.6). Early in fig development the wall of the fig is thin and wasp species that lay their eggs then have short ovipositors. Later in fig development, the wall of the fig is thicker and wasp species that lay their eggs at this later stage have longer ovipositors that are capable of piercing the thick wall of the fig fruit.

The number of duplication events in a given system is constrained by the number of parasites that can coexist on a single host species without competitive exclusion (chapter 6). For example, it is unusual to observe more than one species of a given genus of feather louse on a single species of bird. Most birds have two to four species of lice, but these usually belong to different genera that partition microhabitats. This fact suggests that duplication events have not been very common in the history of feather lice. This kind of logic comes into play when evaluating whether a host switch or duplication event has been more likely when interpreting incongruent phylogenies between hosts and parasites.

To summarize, duplication should be most likely in parasites with patchy distributions, or where there are unfilled microhabitats on the host. All else being equal, closely related parasites that show evidence of competitive exclusion on the same host are unlikely to be associated with parasite duplication events.

Sorting events (extinction and "missing the boat")

Sorting events occur when a parasite lineage goes extinct on a host lineage (fig. 9.1g), or when a parasite fails to disperse with one host lineage leading to founder effect speciation (also called "missing the boat") (fig. 9.1h). Like duplication, the phrase "sorting event" is taken from analogous cases

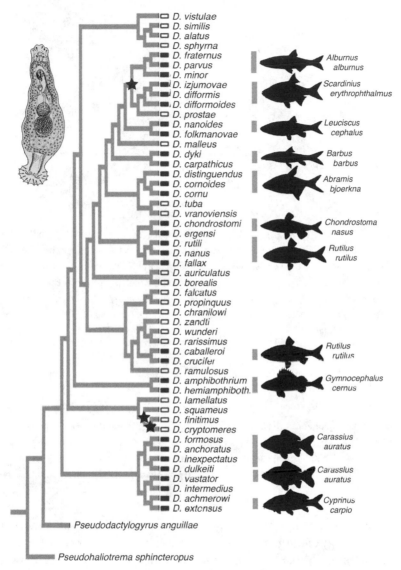

FIGURE 9.5. Phylogeny of *Dactylogyrus* species (Mongenea), showing reconstructed cospeciation events (stars), speciation by host switching (open rectangles), and intra-host duplications (filled rectangles). Fish silhouettes are the hosts on which duplications are inferred to have taken place. After Guilhem et al. (2012).

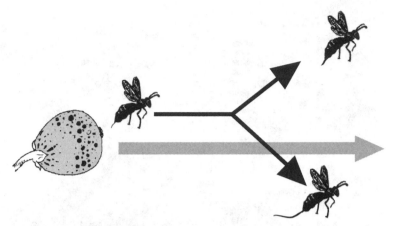

FIGURE 9.6. Probable duplication by parasitic fig wasps; black arrows show speciation, with time indicated by the grey arrow. Both species use their ovipositors to pierce the wall of figs and lay eggs inside. The top species lays its eggs early in fig development, while the wall of the fig is thin. The bottom species lays its eggs late in fig development, after the wall of the fig has thickened considerably. Note that the two species have different-length ovipositors, suggesting ecological speciation on the single species of fig. Drawing based on data from Weiblen and Bush (2002).

FIGURE 9.7. Cophylogenetic reconstruction of wing lice (right) and their Australasian pigeon hosts (left). The New Zealand pigeon (*Hemiphaga novaeseelandiae*) does not have any wing lice associated with it, probably because of a "missing the boat" sorting event (indicated by X) when the ancestor of this pigeon colonized New Zealand. After Johnson and Clayton (2004).

involving gene trees in which a duplicated gene is lost from a genome, making the topology of the gene tree different from that of the species tree (Page and Charleston 1998). Sorting events contribute to incongruence between host and parasite phylogenies, and they reduce parasite diversity. Sorting events increase host specificity if a multi-host parasite goes extinct on one of the species it parasitizes. Australasian wing lice (*Columbicola*) provide a possible example of a sorting event: the New Zealand pigeon (*Hemiphaga novaeseelandiae*) has no wing lice, even though its closest relatives have them (fig. 9.7). Wing lice presumably "missed the boat" when the ancestor of the New Zealand pigeons colonized New Zealand.

Sorting events are most likely to occur when the prevalence and intensity of parasites are low, increasing the probability of stochastic extinction.

Patchily distributed parasites are also more likely to "miss the boat" since this situation increases the odds of a small founding population of hosts being parasite free. Populations of introduced species, including mollusks, crustaceans, fish, birds, mammals, amphibians, and reptiles, have only about half as many parasite species as the same species in their native ranges (Torchin et al. 2003). Similarly, birds that have colonized or been introduced to New Zealand show evidence of sorting events (Paterson et al. 1999). New Zealand birds have lower parasite diversity than their mainland conspecific populations or sister taxa. This fact suggests that the probability that parasites miss-the-boat is higher on hosts that colonize islands. However, MacLeod et al. (2010) have argued that this pattern is more likely to be the result of parasites having gone extinct soon after their hosts were introduced to islands (MacLeod et al. 2010).

Extinction of lice, which is another type of sorting event, is more likely on small or dwindling host populations (Bush et al. 2013). It is also likely in the case of parasites that are rare. For example, of the two genera of lice found on swifts and swiftlets (Apodidae), *Dennyus* is common and *Eureum* is rare. Tompkins and Clayton (1999) carefully searched 1,381 swiftlets for lice; they found *Dennyus* on 23.3% of the birds, but *Eureum* on just 0.3% of birds. All else being equal, one would expect the *Dennyus* phylogeny to be more congruent than the *Eureum* phylogeny with the shared host phylogeny.

It is hard to know how many rare parasite taxa exist in nature unless hosts are sampled in large numbers. A survey of birds in Peru unearthed a new genus of louse (Price and Clayton 1989) that was found on only 3 of 69 (4.3%) wedge-billed woodcreepers (*Glyphorhyncus spirurus*) (Clayton et al. 1992). Unfortunately, most parasite surveys rarely sample more than 10 host individuals, making the detection of rare parasites unlikely. It is unclear how these rare parasites find mates and avoid extinction. Perhaps they are locally or cyclically abundant. Some genera of lice are intrinsically rare. For example, like the genus *Eureum* discussed above, the genus *Hohorstiella* on pigeons and doves seems to be present at low prevalences, even though the hosts are abundant and have been surveyed extensively (Price et al. 2003).

Coadaptive dynamics may factor into the relative frequency of extinction events. In models of host-parasite population dynamics, populations often cycle, which can lead to parasite extinction (Anderson and May 1978; May and Anderson 1978). Aggregation of parasites (chapter 2) tends to stabilize the dynamics of these models. Therefore, it could be that, even though aggregation increases the probability of "missing the boat" events, it counteracts the probability of parasite extinction. Virulent parasites that reduce the reproductive potential of their hosts also tend to cause unstable population cycles (May and Anderson 1978) and have lower prevalence (May and Anderson 1983). Host-parasite systems with virulent parasites may therefore be more prone to parasite extinction than systems involving benign parasites. If this is true, then systems with virulent parasites may be less likely to have phylogenies that are congruent with host phylogenies, all else being equal.

Cohesion

Parasite *cohesion* occurs when a host speciates, yet the parasite does not speciate (fig. 9.1i). This process has also been called parasite "inertia" (Paterson and Banks 2001), or "failure to speciate" (Johnson et al. 2003; Banks and Paterson 2005; Banks et al. 2006). Here, we use the term "cohesion" as a more objective descriptor of the process. In cases of parasite cohesion, which is the opposite of duplication, parasite diversity is unchanged, yet host-specificity decreases. Because cohesion results in a multi-host parasite, it can be difficult to identify this process when reconciling host and parasite phylogenetic trees. This is because most cophylogenetic reconstruction methods either treat multi-host parasites as a recent host switch, or they do not permit them in the analysis (de Vienne et al. 2013).

From a phylogenetic standpoint, cohesion is apparent when one parasite species occurs on two or more sister species of hosts. For example, the wing louse species *Columbicola theresae* occurs on three closely related species of turtle doves (fig. 9.8). As another example, six species of crested penguins (genus *Eudyptes*) have the same species of louse, *Austrogoniodes cristati* (Paterson and Banks 2001), with no evidence of subspecific differentiation (Banks et al. 2006).

Recent switching of a parasite onto a host whose previous parasite has gone extinct can generate a cophylogenetic pattern that resembles cohesion (fig. 9.9; Banks and Paterson 2005; Banks et al. 2006). These alternatives can potentially be distinguished using parasite population genetic data (Banks and Paterson 2005). If the pattern is partly due to a recent host switch, a population bottleneck may be evident in parasites on the recently colonized

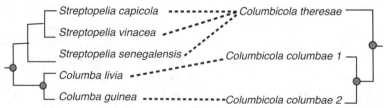

FIGURE 9.8. Cophylogenetic reconstruction of the wing lice (right) of African turtle doves and Old World pigeons (left). *Columbicola theresae* persists as a single species across the three species of African turtle doves (*Streptopelia*), a pattern consistent with parasite cohesion. Bullets are inferred cospeciation events. After Johnson and Clayton (2004).

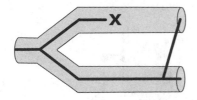

FIGURE 9.9. Parasite extinction (X), followed by a recent host switch involving the parasite's sister species. These events generate a cophylogenetic pattern that resembles parasite cohesion (figure 9.1i). Conventions as in fig. 9.1.

host. Alleles in the population would be a subset of those in the population on the original host species (e.g., Baker et al. 2003). Generally speaking, cohesion is a simpler explanation than the scenario depicted in fig. 9.9, because it does not require two steps: extinction followed by recolonization (Hugot et al. 2001).

Parasite cohesion requires maintenance of parasite dispersal and gene flow between diverging host species. Sympatry of hosts makes such gene flow more likely, yet host speciation is usually associated with allopatric host populations. Cohesion is thus most likely for parasites with higher dispersal potential than their hosts, as in the case of winged herbivorous parasites of plants, or parasites with long-distance vectors.

Pair of lice lost, or parasites regained?

Cophylogenetic methods make it possible to reconstruct the history of host-parasite codiversification (Page and Charleston 1998). Unfortunately, there is often more than one plausible cophylogenetic scenario. Knowledge of the relative timing of host-parasite diversification is often required to distinguish alternative scenarios. A good example involves the sucking lice of the great apes (gorillas, chimpanzees, and humans) and Old World primates (lice have not been collected from orangutans) (Reed et al. 2009; Weiss 2009; Allen et al. 2013). Gorillas and chimpanzees each have a single species of louse, which happen to belong to different genera. *Pthirus gorillae* is found on gorillas, while *Pediculus schaeffi* is found on chimps. In contrast, humans have two species of host specific lice representing both genera: pubic lice (*Pthirus pubis*) and head lice (*Pediculus humanus*). These facts indicate that something other than strict cospeciation must explain the diversity of lice on the great apes.

Comparing the host-parasite phylogenies reveals that each genus of louse is monophyletic, with *Pediculus schaeffi* and *Pediculus humanus* as sister species, and *Pthirus pubis* and *Pthirus gorilla* as sister species (fig. 9.10; Reed et al. 2007). Based on tree topology alone, two scenarios can explain the cophylogenetic pattern: 1) the lice cospeciated with the great apes, then *Pthirus* switched from gorillas to humans and speciated (fig. 9.11a); or 2) lice on the common ancestor of the great apes underwent a duplication event,

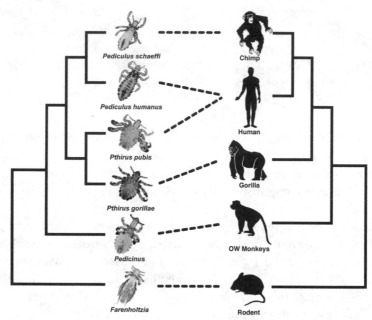

FIGURE 9.10. Cophylogenetic association of primate sucking lice (left) and their hosts (right). Most primates have one species of sucking louse, but humans have two species in different genera. Dashed lines show host-parasite associations. OW = Old World. After Reed et al. (2007).

followed by cospeciation of each lineage, then two extinction events (*Pthirus* going extinct on chimps and *Pediculus* going extinct on gorillas) (fig. 9.11b).

The plausibility of the two scenarios can be compared using information about the relative timing of the different speciation events. In the host switch scenario (fig. 9.11a), speciation of *Pthirus pubis* and *Pthirus gorillae* occurs *after* humans and chimpanzees diverged from their common ancestor. In the duplication/extinction scenario (fig. 9.11b), divergence of *Pthirus pubis* and *Pthirus gorillae* occurs *before* divergence of gorillas and the common ancestor of humans/chimps.

Reed et al. (2007) used DNA sequence data and molecular clock assumptions to reconstruct a time-calibrated phylogeny for these lice. Their study suggested that a combination of the two scenarios might be the most accurate. The split between *Pthirus* and *Pediculus*, at about 13 million years ago, occurred prior to divergence of gorillas and the human/chimp ancestor, estimated at 7 million years ago (fig. 9.12). However, unlike the pure duplication scenario, divergence between *Pthirus gorillae* and *Pthirus pubis* is estimated to have been more recent, at about 3–4 million years ago, suggesting a host switch from gorillas to human ancestors. To reconcile the divergence

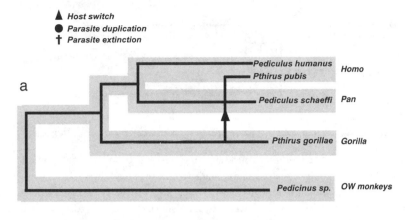

▲ *Host switch*
● *Parasite duplication*
† *Parasite extinction*

a

Pediculus humanus — Homo
Pthirus pubis
Pediculus schaeffi — Pan
Pthirus gorillae — Gorilla
Pedicinus sp. — OW monkeys

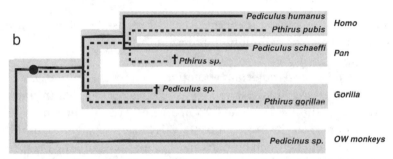

b

Pediculus humanus — Homo
Pthirus pubis
Pediculus schaeffi — Pan
† Pthirus sp.
† Pediculus sp. — Gorilla
Pthirus gorillae
Pedicinus sp. — OW monkeys

FIGURE 9.11. Competing cophylogenetic reconstructions of the great apes (wide grey branches) and lice (thin black branches). (a) Cospeciation events followed by a single host switching event (arrow), (b) Cospeciation with duplication (bullet) into solid and dashed lineages of lice, with two extinctions (crosses). After Reed et al. (2007).

dates with the cophylogenetic patterns, duplication, cospeciation, extinction, and host switching are all required (fig. 9.12).

This reconstruction relies heavily on having relatively accurate calibration dates for the different divergence events in the tree. These dates come from the primate fossil record. Recent fossil evidence (Suwa et al. 2007) of a gorilla-like species from 10–11 million years ago pushes the date for the split between gorillas and the other great apes back further. If true, then a simple host-switching scenario (fig. 9.11a) is the more likely one. In either case, host switching between gorillas and human ancestors is further implied by low genetic divergence between these lice (Allen et al. 2013).

Primate fossils can also be used to provide calibration points for the molecular clock used in reconstructing the evolution of human head and body

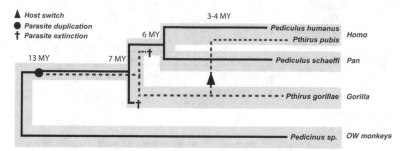

FIGURE 9.12. Cophylogenetic reconstruction of primates and lice, taking into account the results of molecular dating analysis. The scenario includes one duplication (bullet), two extinctions (crosses), and a host switch (arrow) from gorillas to human ancestors. Dates show the approximate timing of these events, as inferred by Reed et al. (2007). After Reed et al. (2007).

lice. The evolutionary history of these lice can be used to shed interesting light on human cultural evolution (box 9.2).

Cophylogenetics and molecular evolution

Cospeciation events link host and parasite lineages together in time. Repeated cospeciation events allow comparisons of evolution in host and parasite groups over the same time intervals (Page 2003). For example, the cospeciation event between humans and chimpanzees and their respective species of *Pediculus* is well supported from phylogenetic topological and molecular dating analyses (Reed et al. 2007; figs. 9.11 and 9.12). Divergence in genes shared by humans and chimps, compared to their common ancestor, can be compared to divergence of the same gene in their lice, compared to its common ancestor. Comparisons of the same mitochondrial genes, for example, reveal that the lice are evolving 2–3 times faster than their primate hosts (Reed et al. 2004).

Elevated molecular genetic substitution rates in lice, compared to hosts, has also been noted in gophers and *Geomydoecus* lice (Hafner et al. 1994), swiftlets and *Dennyus* lice (Page et al. 1998), pigeons and *Columbicola* lice (Johnson et al. 2003), and rodents and *Fahrenholzia* lice (Light and Hafner 2007). These studies indicate that the lice evolve 2–10 times faster than their hosts. All of these studies compared homologous mitochondrial genes, because it is relatively straightforward to sequence the same gene region in both the lice and their vertebrate hosts. In addition to elevated rates of substitution in mitochondrial protein-coding genes, the mitochondrial 12S ribosomal DNA gene shows highly diverged secondary structures (Page et al. 2002). Interestingly, there is more variation in the secondary struc-

BOX 9.2. Using lice to date the origin of clothing

BOX 9.2. Using lice to date the origin of clothing

Rates of molecular evolution in lice can be used to put an approximate date on the cultural evolution of clothing in human ancestors. This approach involves human head lice (*Pediculus humanus capitis*) and body lice (*P. h. humanus*) (chapter 2). These subspecies have subtle morphological differences (Reed et al. 2004), and more dramatic ecological differences. Head lice attach their eggs to head hair, while body lice attach their eggs to clothing. Given these differences, the date of divergence between head and body lice can be used to estimate the minimum amount of time since humans started wearing clothing—that is, the availability of the discrete microhabitat for body lice. On the basis of mitochondrial DNA sequences, Kittler et al. (2003) found little divergence between head and body lice, suggesting a recent date for the origin of clothing, consistent with the first appearance of eyed needles in the archaeological record about 40,000 years ago. However, additional data from nuclear loci, together with Bayesian coalescent modeling, pushes the origin of the body louse back to 83,000–170,000 years ago (Toups et al. 2011).

The *P. humanus* story is actually more complicated than simple divergence between head and body lice. Multiple deeply divergent mitochondrial lineages have been shown within head lice, only one of which is shared with body lice (Kittler et al. 2003; Reed et al. 2004). Estimates of the dates of divergence among these lineages suggest that one of them diverged about 1.2 million years ago, around the time of the divergence between the ancestors of modern humans and Neanderthals. The other lineage is older, dating closer to the divergence between *Homo erectus* and other hominids (Allen et al. 2013). Reed et al. (2004) suggested that the *H. erectus* and Neanderthal lice lineages survive today because they switched onto the lineage leading to modern humans, thus avoiding coextinction with their original hosts. On the other hand, the same divergence is not evident in the nuclear DNA data for the lice (Kittler et al. 2003). Thus, it is possible that the deeply divergent lineages in mitochondrial genes of human head lice represent a lack of sorting of ancestral polymorphisms, given the rapid historical population expansion of humans and, by inference, their head lice.

tures (stems and loops) of this gene across lice, than across all other insects combined.

Mitochondrial gene order for lice is also highly rearranged compared to the relatively conserved gene order across other orders of insects. In the case of the human body louse (*Pediculus h. humanus*), the mitochondrial genome is actually divided across several minicircular chromosomes (Kirkness et al. 2010). Often only a single protein-coding gene is found on each of these minicircles. Other species of lice also have mitochondrial minicircles or smaller chromosomes containing a subset of the usual mitochondrial protein-coding genes (Cameron et al. 2011).

It is not yet clear whether the elevated substitution rates seen in lice are somehow connected to the unusual mitochondrial gene rearrangements and minicircles in lice. One possibility is that the human body louse genome lacks mitochondrial single-stranded binding protein (mtSSB), which helps stabilize mitochondrial DNA and protect it from mutation while it is in its single-stranded state during replication (Cameron et al. 2011). Lack of this protein might be linked to both the formation of minicircles and elevated mutation rates. Relaxed selection or accumulation of nonsense deleterious mutations in this gene might be an explanation for the loss of mtSSB in the first place.

These hypotheses assume that elevated rates of molecular evolution occur only in the mitochondrial genes of lice. However, the recent affordability of DNA sequencing now makes it possible to use cophylogenetic approaches with a variety of molecular markers, or even complete genomes. Johnson et al. (2014) examined a set of 1,534 nuclear protein-coding genes for the human-chimp and *Pediculus* spp. cospeciation event. The authors sequenced the genome of the chimp louse (*Pediculus schaeffi*) and compared it to existing genomes for humans, chimps, and the human body louse (*Pediculus h. humanus*). They found that every gene in the lice was evolving faster than the respective gene in the hosts (fig. 9.13). On average, the genes in lice were evolving 15 times faster than those in the hosts. Thus, the elevated rate of molecular evolution in lice is definitely not restricted to mitochondrial genomes.

Johnson et al. (2014) also found a correlation between the amount of divergence for a particular gene between humans and chimps, and the amount of divergence in the same gene for their lice (fig. 9.13). That is, genes that have evolved more rapidly in humans and chimps tend to have also evolved more rapidly in their lice. This correlation occurred only in the case of mutations that change amino acids, and thus possibly the function of the protein, suggesting that genes that are highly constrained in primates are also highly constrained in their lice. Across all genes, the ratio of substitutions that altered an amino acid to substitutions that did not alter them was significantly higher in primates than in their lice. This result is consistent with the lower effective population size of primates, compared to their lice, which could allow more amino acid alterations to go to fixation by random genetic drift, even if they are detrimental (Ohta 1992; box 8.1). Thus, increased substitution rates by bottleneck effects in lice are not likely to explain their faster rates of molecular evolution (cf. Page et al. 1998).

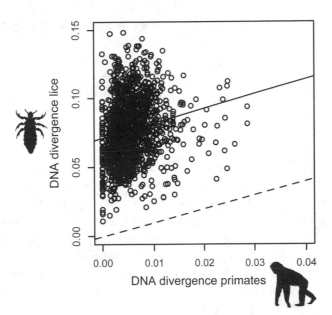

FIGURE 9.13. Plot of pairwise uncorrected sequence divergence for 1,534 orthologous protein-coding genes between humans and chimpanzees (x-axis) and their lice (y-axis). Solid line is least squares regression (slope = 1.08, P < 0.0001). Dashed line is the expectation if genes evolve at same rate in the two groups. Points above the line are genes that evolve faster in lice. Points below the line (none in this case) are genes that evolve faster in primates. After Johnson et al. (2013).

Two other factors might explain the faster rates of molecular evolution in lice, compared to hosts. First, lice have much shorter generation times than their hosts, which may provide more opportunity for mutations per unit time. Secondly, because they are parasites, selection across the genome might be relaxed, because some genes, such as those involved in the production of wings or complex eyes, are no longer needed. Genome-wide relaxed selection may also lead to relaxed selection for mutation repair (Leigh 1970; Yoshizawa and Johnson 2013). Either of these mechanisms, or some combination of them, seems plausible. More detailed investigation of mutation repair enzymes for nuclear genes could help distinguish the two mechanisms.

In summary, this chapter has tried to illustrate the power of cophylogenetic approaches. Cospeciation, host switching, duplication, sorting events, and cohesion are the processes governing patterns of codiversification between parasites and their hosts. Depending on the ecological underpinnings of a given system, one or more of these processes may dominate. The challenge is to understand which processes are favored by different ecological circumstances.

There were no lice in the Garden of Eden.
Such loathsome creatures must have been created after the Fall.
—Rothschild and Clay 1957

Cophylogenetic patterns are central to the study of co-evolution in the broad sense because they document codiversification, which is the correlated diversification of interacting lineages (box 1.1). In the last chapter, we reviewed the macroevolutionary events that govern the cophylogenetic dynamics of codiversifying groups. Given the complexity of these processes, it can be difficult to identify ecological or other factors influencing macroevolutionary events and cophylogenetic processes. Comparisons of related groups of organisms can be very helpful. This comparative approach has the power to pinpoint ecological or other differences between groups that may be responsible for different patterns of codiversification.

In this chapter we apply this comparative cophylogenetic approach to groups of lice with different cophylogenetic histories. We examine the influence of cospeciation and host switching on patterns of codiversification. Cospeciation and host switching are the ends of a continuum with opposite effects on cophylogenetic patterns. Both processes contribute to codiversification, but in different ways. Lice span the cospeciation–host switching continuum, making it feasible to explore the relationship of ecology to cophylogenetic history within a single insect order. This, in turn, reduces the likelihood of drawing erroneous conclusions from comparing apples and oranges that have evolved in completely different ecological contexts. The case studies reviewed below illustrate some of the ways in which to integrate cophylogenetic analysis with ecological data at the interface of comparative and experimental approaches. They also illustrate the critical importance of understanding the basic natural history of the system being studied.

Extensive cospeciation: Pocket gopher lice

Pocket gophers and their lice show extensive cophylogenetic congruence (fig. 10.1; Hafner et al. 2003). Pocket gophers (Rodentia: Geomyidae) dig elaborate tunnel systems and spend most of their lives underground. They are solitary and highly territorial, rarely contacting other individuals of their own species, let alone other species. Pocket gophers are host to lice in

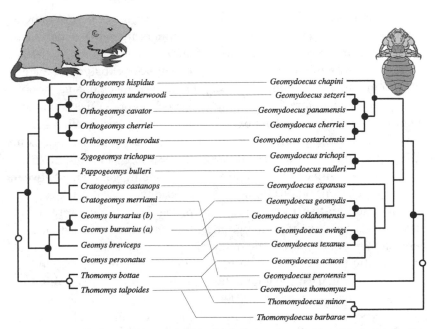

FIGURE 10.1. Cophylogenetic reconstruction of pocket gophers (left) and their chewing lice (right) showing extensive cospeciation. Black bullets are inferred cospeciation events within *Geomydoecus*. White bullets are inferred cospeciation events within *Thomomydoecus*. Thin lines show host-parasite associations. After Hafner et al. (2003).

two ischnoceran genera: *Geomydoecus* and *Thomomydoecus*. Species in these genera are extremely host-specific, often being found on a single species or even subspecies of gopher (Price et al. 2003). While the two genera often coexist on the same host, it is rare to find two species of the same genus on the same host. When they do co-occur, *Geomydoecus* and *Thomomydoecus* partition the host's body, with *Geomydoecus* preferring dorsal surfaces and *Thomomydoecus* preferring ventral surfaces (fig. 6.2; Reed et al. 2000b). Competition may play a role in this microhabitat partitioning, although this hypothesis needs to be tested experimentally (see chapter 6).

Like all lice, gopher lice spend their entire life cycle on the body of the host, gluing their eggs to hairs and feeding on sloughed skin. While most transmission of these lice is thought to be through vertical transmission from mother to offspring, genetic evidence suggests that other transmission routes may also be important (Demastes et al. 1998, 2012). Some transmission may occur during male-female contact, or in territorial disputes. However, the relatively tight linkage of gopher louse reproduction and dispersal with host reproduction and dispersal sets the stage for frequent bouts of cospeciation, leading to cophylogenetic congruence.

What about other processes that may have played a role in the cophylogenetic history of gophers and lice? Pocket gophers are highly allopatric in distribution, and even when the ranges of different gopher species overlap, syntopy (shared habitat) is rare. Gopher lice are not capable of efficient locomotion off the host's body, and they do not have mechanisms for phoretic dispersal. Therefore, host switching by pocket gopher lice is unlikely. The probability of maintaining parasite gene flow between pocket gopher species is also low, although gene flow in lice across a gopher hybrid zone has been documented (Nadler et al. 1990; Hafner et al. 1998). Similarly, parasite cohesion is unlikely because it requires ongoing gene flow between parasite populations on diverging host populations.

Duplication and sorting events have probably also been rare in the coevolutionary history of pocket gophers and their lice. Because the host spends its life underground, the microclimate is very stable (Hafner et al. 2003). Abiotic factors, such as humidity, are relatively constant and no major distributional gaps in gopher lice have been observed. This relatively continuous distribution of lice presumably makes duplication events rare. Gopher lice show high prevalence, reaching 100% in many populations. The intensity of lice on pocket gophers is also high, with most individual hosts harboring more than 100 lice. The high prevalence and intensity of gopher lice makes sorting events unlikely, either through extinction or "missing the boat."

Preliminary experimental transfers show that pocket gopher lice have the ability to survive and reproduce on novel hosts that are closely related to the native host (Reed and Hafner 1997). This ability is probably due to the fact that related hosts are similar in size, facilitating attachment to hairs with the mandibles and rostral groove (fig. 5.3; Hafner et al. 2003). For a successful host switch to occur under natural conditions, however, both dispersal and establishment must be feasible. Limitations to dispersal appear to have made host switching a rare event in the history of pocket gopher lice, despite the apparent potential for establishment on some novel hosts.

In summary, cospeciation by gopher lice with hosts is expected to be common, with few events expected to erode the congruence between the host and parasite trees. In support of this prediction, Hafner et al. (1994) documented extensive cophylogenetic congruence between pocket gophers and *Geomydoecus* lice (fig. 10.1). Their cophylogenetic reconstruction inferred that 8 of 12 (67%) speciation events in pocket gophers have a corresponding speciation event in the lice, which is many more than expected by chance (Hafner et al. 2003; Clayton et al. 2004). Although the study by Hafner et al. (1994) was at a higher taxonomic scale, cophylogenetic studies of pocket gopher lice at finer scales, even among subspecies, also reveal

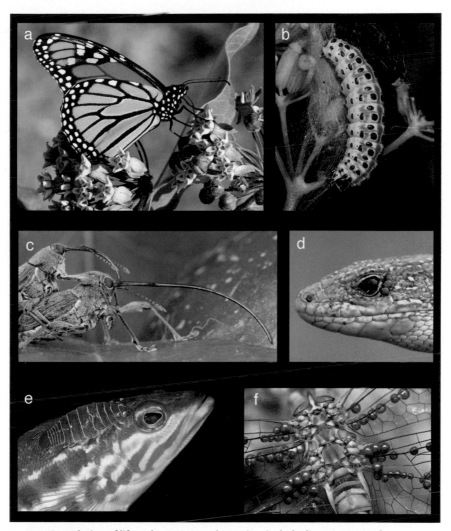

PLATE 1. Coevolution of life on hosts. External parasites include diverse groups of arthropods on plant or animal hosts. Coevolution occurs between phytophagous insects and plants, such as (a) monarch butterflies (*Danaus plexippus*) and milkweeds (*Asclepias syriaca*) (by B. Hysell); (b) parsnip webworms (*Depressaria pastinacella*) and wild parsnips (*Pastinaca sativa*) (by Nicolas Moulin Photography); and (c) camellia weevils (*Curculio camelliae*) and Japanese camellia (*Camellia japonica*) (by H. Toju). Similarly, coevolution takes place between ectoparasites and animal hosts, such as (d) ixodid ticks and green lizards (*Lacerta viridis*) (by J. Niedobová); (e) parasitic isopods (*Anilocra physodes*) and comber fish (*Serranus cabrilla*) (by L. A. Díaz); and (f) water mites (*Arrenurus* sp.) and dragonflies (by P. Ambruzs).

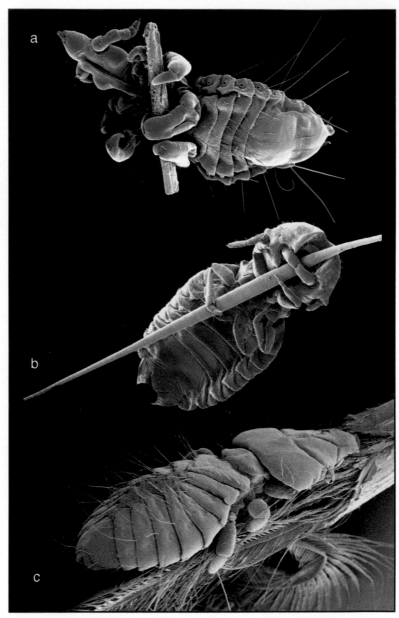

PLATE 2. Colorized SEMS of lice (Phthiraptera: Ischnocera); (a) Sucking louse *Linognathus euchore* (Linognathidae) using three legs to cling to the hair of an African springbok (*Antidorcas marsupialis*) (SEM by M. Turner, E. Green, and J. Wentzel); (b) Mammal chewing louse *Damalinia ornata* (Trichodectidae) using both mandibles and two legs to grasp the hair of an African hartebeest (*Alcelaphus buselaphus*)—note the hair is also passing through the rostral (ventral) head groove (SEM by M. Turner); (c) Avian chewing louse *Saemundssonia lari* (Philopteridae) on the feather of a grey-headed gull (*Larus cirrocephalus*) (SEM by M. Turner and F. C. Clarke).

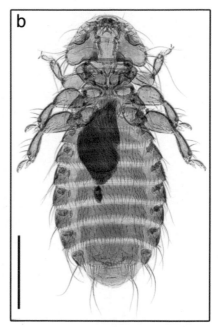

PLATE 3. Fossil bird louse *Megamenopon rasnitsyni*; (a) Complete exoskeleton; scale bar = 2 mm; (b) Extant species of *Holomenopon brevithoracicum* (Menoponidae) from a mute swan (*Cygnus olor*) with feather barbules in its crop (the large reddish brown structure); scale bar = 0.5 mm. Preserved feather barbules can also be seen upon microscopic examination of the fossil louse. Reprinted from Wappler et al. (2004), by permission of the Royal Society.

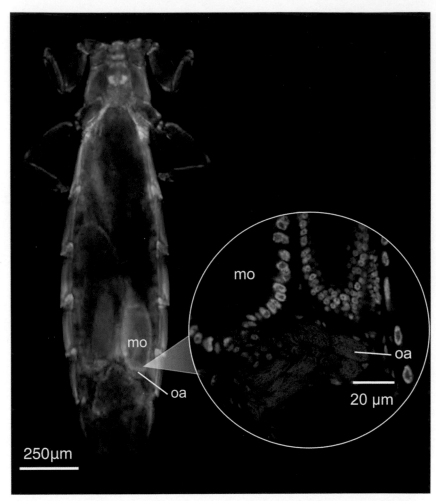

PLATE 4. Bacteriocytes in the abdomen of an adult female *Columbicola columbae* (left). Red and green colors show bacterial and louse cells, respectively. The bacteriocytes form conspicuous tissues called ovarial ampullae (oa) that are associated with developing eggs (mature oocytes: mo). Inset shows vertical transmission, with bacterial cells moving from the ovarial ampulla to the posterior pole of an oocyte through follicle cells (adapted from Fukatsu et al. 2007). See box 2.1 for further details.

PLATE 5. (a) Hippoboscid fly collected from a Eurasian blackbird (*Turdus merula*) with colorized phoretic lice (*Brueelia* sp.). Inset: enlarged view of the attached *Brueelia*. SEM by V. S. Smith; specimen collected by R. Dawson in Agadir, Morocco. (b) Pigeon fly (*Pseudolynchia canariensis*) with colorized phoretic wing lice (*Columbicola columbae*). Inset: enlarged view showing *C. columbae* using its third pair of long legs to grasp the fly's leg (SEM by E. H. Burtt Jr. and J. Ichida, adapted from Harbison and Clayton 2011).

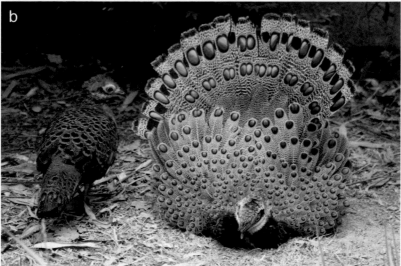

PLATE 6. Lice and other parasites mediate sexual selection for host traits that reveal freedom from parasites and pathogens. Species with more ornamental plumage, such as the (a) Indian peafowl (*Pavo cristatus*), must devote more time to maintenance behavior than related species with less ornamental plumage, such as the (b) Bornean peacock pheasant (*Polyplectron schleiermacheri*). Hence, ornamental plumage has a cost in terms of the time and energy required to keep it in good condition, which is the basis of the "high maintenance" handicap principle. Photos by (a) by A. Murray, naturepl.com, and (b) by O. Taylor, World Pheasant Association.

PLATE 7. (a) Allogrooming by Japanese macaques (*Macaca fuscata*); (b) Allogrooming by Waorani (Huaorani) women and children from Ecuador. Mutual grooming is common in other groups of primates, and other human societies. Photos by (a) Noneotuho, Wikimedia; and (b) Broennimann (1981).

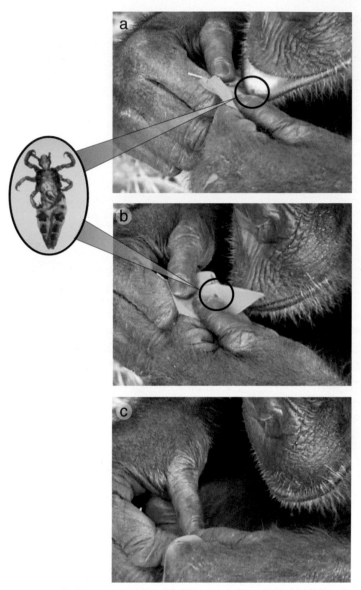

PLATE 8. Stills from a video of a wild chimpanzee (*Pan troglodytes*) filmed as it transferred an allogroomed louse from its lower lip (a) to a leaf (b), which it then folded (c), crushing the louse, which was subsequently recovered (inset) from the discarded leaf. Thus, chimpanzees are capable of using leaves as grooming aids in the control of lice (from Zamma 2002b).

PLATE 9. (a) Preening spotted dove (*Streptopelia chinensis*); (b) Allopreening hooded vultures (*Necrosyrtes monachus*). Preening is an effective first line of defense against bird lice; allopreening may also help control lice. Photos by (a) J. M. Garg, Wikimedia, and (b) Dave Montreuil, fotolia.com.

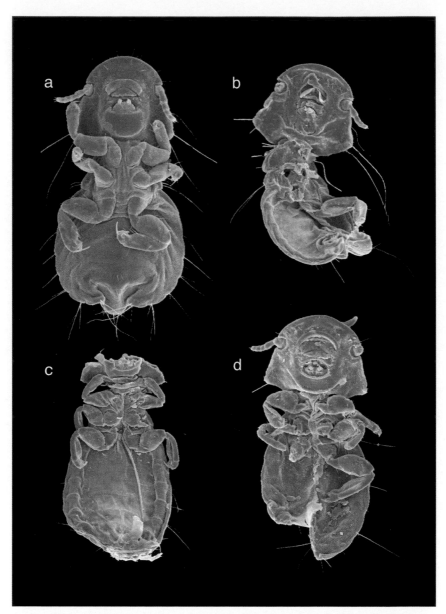

PLATE 10. Intact body louse (*Campanulotes compar*) from a rock pigeon (a), compared to body lice with missing legs (b), decapitation (c), or traumatic damage to the abdomen (d). More lice suffer this kind of damage on birds with intact overhangs, compared to birds from which the overhang has been trimmed. Colorized SEMs adapted from Clayton et al. (2005).

PLATE 11. Dusting and sunning behavior: (a) Dust-tossing by an African elephant (*Loxodonta sp.*); (b) Dusting by a southern ground-hornbill (*Bucorvus leadbeateri*); (c) Sunning by a European bee-eater (*Merops apiaster*). Photos by (a) Dreamstime LLC, (b) T. Laman, naturepl. com, and (c) K. Wothe, naturepl.com.

PLATE 12. Cryptic coloration in bird lice. The light-colored louse, *Neopsittaconirmus albus*, parasitizes the sulfur-crested cockatoo (*Cacatua galerita*) (a). The dark-colored louse, *Neopsittaconirmus borgiolii*, parasitizes the yellow-tailed black cockatoo (*Calyptorhynchus funereus*) (b). Host feathers are the natural background for these lice. Both species of lice were photographed on feathers from a sulfur-crested cockatoo (c) and a yellow-tailed black cockatoo (d). Figure adapted from Bush et al. (2010); cockatoo photos by (a) JJ Harrison, Wikimedia, and (b) D. Cook, Wikimedia.

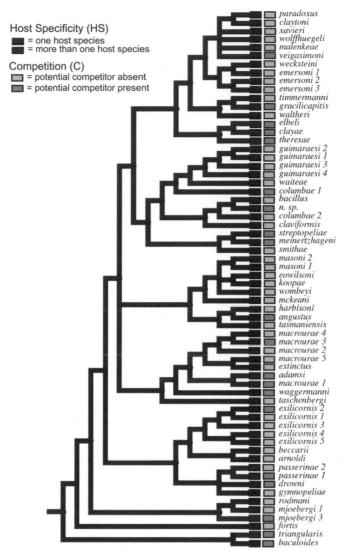

Host Specificity (HS)
■ = one host species
■ = more than one host species

Competition (C)
▢ = potential competitor absent
▢ = potential competitor present

paradoxus
claytoni
xavieri
wolffhuegeli
malenkeae
veigasimoni
wecksteini
emersoni 1
emersoni 2
emersoni 3
timmermanni
gracilicapitis
waltheri
elbeli
clayae
theresae
guimaraesi 2
guimaraesi 1
guimaraesi 3
guimaraesi 4
waiteae
columbae 1
bacillus
n. sp.
columbae 2
claviformis
streptopeliae
meinertzhageni
smithae
masoni 2
masoni 1
eowilsoni
koopae
wombeyi
mckeani
harbisoni
angustus
tasmaniensis
macrourae 4
macrourae 3
macrourae 2
macrourae 5
extinctus
adamsi
macrourae 1
waggermanni
taschenbergi
exilicornis 2
exilicornis 1
exilicornis 3
exilicornis 4
exilicornis 5
beccarii
arnoldi
passerinae 2
passerinae 1
drowni
gymnopeliae
rodmani
mjoebergi 1
mjoebergi 3
fortis
triangularis
baculoides

PLATE 13. Evolution of host specificity (first column), relative to the presence of potential competitors (second column). Parsimony reconstruction of host specificity (first column: host specialist = red; host generalist = dark blue) over strict consensus parsimony tree (Johnson et al. 2007) for species of dove wing lice (*Columbicola*). Over the tree there are six independent gains of generalists from host-specific ancestors. Tiny pie charts on each branch are marginal probabilities, from maximum likelihood, of the ancestral state being a specialist (red) or a generalist (dark blue). These gains of host generalization were highly correlated with the presence of a potentially competing species of *Columbicola* (second column: no other species present = pink, another species present = light blue) on at least one of the host species (concentrated changes test [Maddison 1990], P < 0.003 for parsimony and Bayesian trees, maximum likelihood correlation [Pagel 1994], P < 0.01 over both trees). Reprinted from Johnson et al. (2009) by permission of the Royal Society.

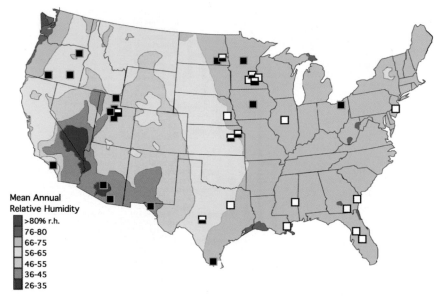

PLATE 14. Distribution of two species of *Columbicola* on mourning doves in the United States in relation to mean annual relative humidity (NOAA 2002). Black squares are collecting localities for *C. baculoides*; white squares are collecting localities for *C. macrourae*; two-tone squares are localities where *C. baculoides* and *C. macrourae* co-occur, sometimes on the same individual dove. After Malenke et al. (2011).

(*opposite*)

PLATE 15. Galápagos hawks (*Buteo galapagoensis*) and their parasitic lice vary in population genetic structure. (a) STRUCTURE plots (K = 8) estimating the most probable number of genetic clusters for hawks and their lice (*Degeeriella regalis*). Each thin vertical line represents an individual hawk or louse, with the island from which they were sampled shown at the top. Colors correspond to the common haplotypes on the color-coded islands (b) (after Koop et al. 2014). (c) Population structure estimated by mitochondrial DNA haplotype networks for Galápagos hawks and their three species of lice: *Degeeriella regalis*, *Colpocephalum turbinatum*, and *Craspedorrhynchus* sp. Lines connect haplotypes by single mutational steps. Small black bullets are inferred haplotypes (unsampled or extinct). Observed haplotypes are indicated by circles and squares, with the size proportional to the number of individuals sampled. Colors correspond to the map and show the geographic origins of sampled individuals. Squares represent the putative oldest haplotypes. After Whiteman et al. (2007, 2009).

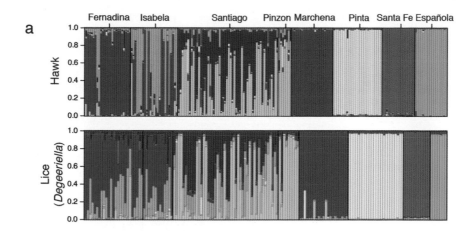

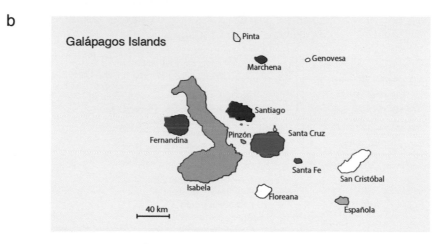

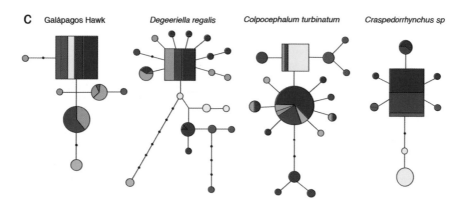

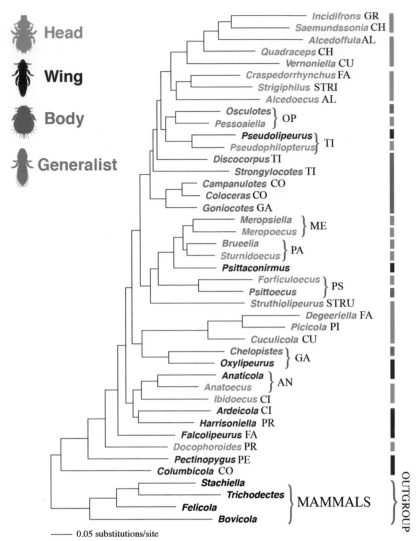

PLATE 16. Molecular phylogeny of feather lice (Ischnocera) with representatives of different ecomorphs indicated by different-colored font and vertical bars. Brackets highlight cases in which the ecomorphs on a single host group are sister taxa. Hosts are designated by the following letter codes: AL, Alcedinidae; AN, Anseriformes; CH, Charadriiformes; CI, Ciconiiformes; CO, Columbiformes; CU, Cuculiformes; FA, Falconiformes; GA, Galliformes; GR, Gruiformes; ME, Meropidae; OP, Opisthocomidae; PA, Passeriformes; PE, Pelecaniformes; PI, Piciformes; PR, Procellariiformes; PS, Psittaciformes; STRI, Strigiformes; STRU, Struthioniformes; TI, Tinamiformes. From Johnson et al. (2012).

significant congruence between pocket gopher and louse phylogenies (De-mastes et al. 2003, 2012; Light and Hafner 2007).

Moderate cospeciation with some host switching: Swiftlet lice

Lice in the genus *Dennyus* show significant cophylogenetic congruence with their hosts, which are swifts and swiftlets (Apodidae). However, they also show evidence for events that erode congruence, such as host switching, duplication, and sorting events (fig. 10.2). What ecological conditions may have contributed to these events?

Swiftlets are distributed throughout Australasia, with many endemic species on islands in the Indian and South Pacific oceans (Lee et al. 1996). Thus, the biogeography of swiftlets is patchwork, with strong allopatric isolation (Clayton et al. 2003a). This pattern creates the potential for duplication, which may explain cases in which more than one species of *Dennyus* occur on single host species (Clayton et al. 1996). *Dennyus* prevalence is often high (Lee and Clayton 1995), but intensity tends to be low, with fewer than ten lice on most birds (Tompkins and Clayton 1999). These factors create the potential for sorting events. Interestingly, some island populations of swiftlets, such as *Aerodramus bartshii* (Hawaii), and *A. sawtelli* (Cook Islands),

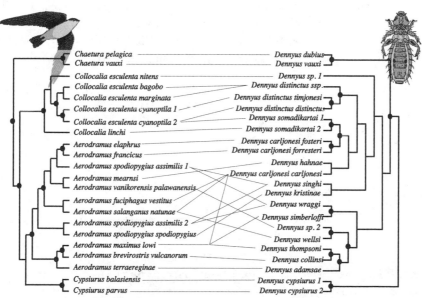

FIGURE 10.2. Cophylogenetic reconstruction of swifts and swiftlets (left) and *Dennyus* lice (right) showing moderate cospeciation with some host switching. Numbers after the names of some lice indicate cryptic species based on molecular work. Black bullets are inferred cospeciation events. Thin lines show host-parasite associations. After Clayton et al. (2003a).

appear to be free of *Dennyus*, despite concerted sampling efforts (unpublished data).

The allopatric distribution of most species of swiftlets makes host switching and cohesive gene flow between host species unlikely. However, not all species of swiftlets are allopatric. In some locations, up to four species of cave-dwelling swiftlets can be found nesting in proximity to one another. Although transmission of *Dennyus* is largely vertical (fig. 7.2), there may be some horizontal dispersal between nests. Tompkins and Clayton (1999) censused *Dennyus* from 1,381 swiftlets belonging to four species nesting in a single cave in Borneo. The prevalence of a given species of *Dennyus* tended to be very high on some swiftlet species, but extremely low (<5%) on other species in the same cave. This pattern suggests that, like other Amblycera, *Dennyus* is capable of trickle dispersal between host species nesting in close proximity. Phoretic dispersal is virtually unknown in Amblycera (chapter 7), and not expected in this case because the hippoboscid flies on swifts are wingless (Tompkins et al. 1996).

Can *Dennyus* survive on novel host species if they disperse to them? Tompkins and Clayton (1999) conducted field transfer experiments in which they moved *Dennyus* between species of swiftlets. Transfers included those between host species on which non-specific *Dennyus* are normally found, as well as transfers to novel host species on which they are not normally found. The lice were able to survive on both native and novel hosts, so long as the wing feather barbs of the host species did not differ much in size (fig. 10.3; Tompkins and Clayton 1999). The importance of size in this case

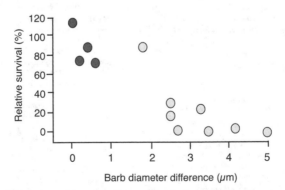

FIGURE 10.3. Relative survival (compared to controls) of swiftlet lice (*Dennyus*) transferred between host species, compared to the difference in feather barb diameter of the donor and recipient host species. Dark grey points are transfers of non-specific lice between hosts on which they normally occur; light grey points are transfers of lice from native to novel host species. Survival was inversely correlated to the barb diameter difference (P < 0.001). After Tompkins and Clayton (1999).

was thought to reflect the ability of *Dennyus* to remain attached to hosts when they flap their wings. Interestingly, even on some hosts with different feather barb sizes, transferred lice could hold feather size "constant" by altering their microhabitat distributions along the wings. In short, behavioral plasticity allowed the lice to survive on host species over a broader range of sizes than would have been possible otherwise.

Given the limited dispersal capabilities of *Dennyus*, one might predict a history of extensive host-parasite cospeciation. However, as discussed above, there are also opportunities for duplication, sorting events, and host switching. Cophylogenetic analyses of swiftlets and *Dennyus* (Page et al. 1998; Clayton et al. 2003a) reveal that 12 out of 20 (60%) host speciation events are associated with *Dennyus* cospeciation events (fig. 10.2). This is a smaller fraction than that observed for the gopher-*Geomydoecus* system, yet still more than expected by chance (P < 0.01).

Extensive host switching: Songbirds and allies and their lice

Lice in the genus *Brueelia* occur on songbirds (Passeriformes) and four other related orders of birds: Coraciiformes (bee-eaters), Cuculiformes (cuckoos), Piciformes (woodpeckers), and Trogoniformes (trogons). *Brueelia* is one of the most diverse genera of bird lice, with several hundred species, most of which are host specific (Price et al. 2003). Despite their specificity, however, there is little correspondence between the *Brueelia* and host phylogenies; cospeciation is relatively uncommon in this group (fig. 10.4). Ecological considerations suggest that host switching and other events are more likely in *Brueelia* than in the pocket gopher or swiftlet lice considered above. Why the difference?

Brueelia occurs mainly on small-bodied birds and, as such, parasite loads are normally low. Prevalence is often less than 10%, and intensity is usually less than 10 lice per infested bird (Hahn et al. 2000; Najer et al. 2012). Low parasite loads increase the probability of sorting events, such as extinction and "missing the boat." The geographic distribution of *Brueelia* can also be restricted. Little is known about the geographic distributions of most *Brueelia*; however, a study by Bush et al. (2009a) showed that the range of *Brueelia deficiens* on Western scrub jays (*Aphelocoma californica*) is smaller than the range of the host (fig. 10.5). The louse is restricted to arid regions, possibly as a refuge from competition with *Myrsidea* sp.

Condition-dependent competition of this type explains limited geographic distributions in lice on mourning doves (plate 14; chapter 6). *B. deficiens* is also a parasite of pinyon jays (*Gymnorhinus cyanocephalus*), which are restricted to arid regions, suggesting that they may serve as a reservoir. Discontinuous parasite distributions can increase the likelihood of parasite

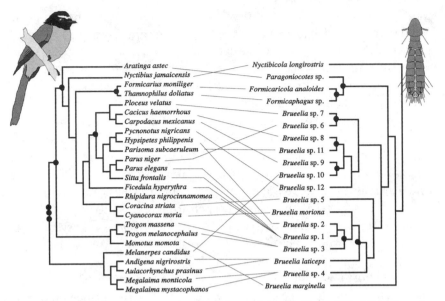

FIGURE 10.4. Cophylogenetic reconstruction of songbirds and allies (left) and their *Brueelia* lice (right). The analysis shows little cospeciation, but frequent host switching. The number of cospeciation events is not more than would be expected by chance. Conventions as in fig. 10.2. After Johnson et al. (2002a).

duplication, which could explain why some species of songbirds are host to more than one species of *Brueelia*. Uneven distributions also increase the potential for sorting events.

Brueelia also shows considerable potential for host switching and cohesion. Songbird species have strongly overlapping distributions, with more than 100 sympatric species occurring in the same geographic region. Many hosts of *Brueelia* are hole-nesters (Johnson et al. 2002a), and limited numbers of holes can lead to competition and interspecific nest takeovers (Merilä and Wiggin 1995). Since *Brueelia* can survive off the host, at least for short periods of time (Dumbacher 1999), nest hole disputes may facilitate host switching.

Brueelia is also the most commonly phoretic genus of louse (Keirans 1975a). Roughly 80% of 350 records of phoresis on hippoboscid flies are passerine lice, with the majority being the genus *Brueelia*, or the closely related genus *Sturnidoecus*. Phoresis by *Sturnidoecus* on European songbirds has been recorded at high frequencies (table 7.1). *Brueelia* appears to be capable of establishing populations on a wide variety of hosts; indeed, some species of *Brueelia* share hosts that belong to different taxonomic families of birds (Johnson et al. 2002b). *Brueelia* has been described as a microhab-

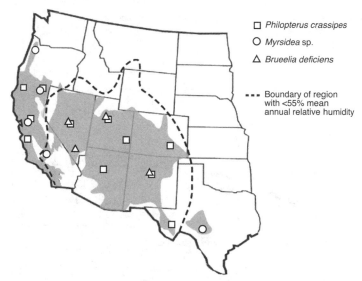

FIGURE 10.5. Distribution of the western scrub jay (*Aphelocoma californica*) (grey) and its lice (see legend). *Myrsidea* is mostly absent from arid regions (inside dotted line), while *Brueelia* is absent from relatively humid regions (outside dotted line). After Bush et al. (2009a).

itat generalist that roams over much of the host's body (Clay 1951). This behavioral plasticity may help *Brueelia* survive and reproduce on hosts of different sizes.

In conclusion, ecological considerations suggest that duplication, sorting events, host switching, and cohesion may all have been common events in the macroevolutionary history of *Brueelia*. Not surprisingly, therefore, a study comparing the phylogeny of 15 species of *Brueelia* to that of their hosts (Johnson et al. 2002a) found no evidence for cophylogenetic congruence (P = 0.25) (fig. 10.4). Only 5 of 24 (20%) host speciation events had inferred cospeciation events by lice, which is not more than expected by chance (Clayton et al. 2004). Finally, recent taxonomic work suggests that the apparent host specificity of *Brueelia* may be partly due to oversplitting on the basis of host associations, rather than on the basis of the lice themselves (unpublished data). This problem has plagued the taxonomic history of lice, in general (Price et al. 2003).

Major host switching: Flamingo lice

Lice occasionally also show evidence of major host switching events between unrelated hosts—for example, the members of different host orders, or even classes (table 2.1). In such cases, the novel host may be rela-

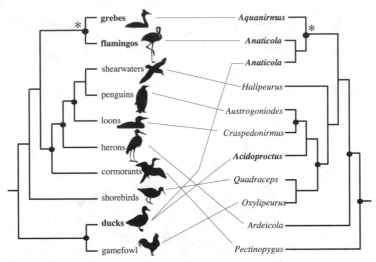

FIGURE 10.6. Cophylogenetic reconstruction of aquatic bird orders and their genera of lice. Nodes with asterisks represent a cospeciation event inferred between the ancestor of flamingos and grebes, and the ancestor of the lice genera *Aquanirmus* and *Anaticola*. A major host switch by a member of the *Anaticola* to ducks was followed by extensive diversification of this genus among waterfowl (see text). Conventions as in fig. 10.2. After Johnson et al. (2006).

tively defenseless, facilitating establishment of the lice. Establishment of novel lice could conceivably also be facilitated by release from competition with more closely related lice (chapter 6). To our knowledge, however, neither the "defenseless host" nor the "competitive release" hypotheses have been tested for lice.

An interesting example of a major host switch, followed by extensive diversification, involves the lice of flamingos, ducks, and other aquatic birds (fig. 10.6). Flamingos share three genera of lice with ducks and other waterfowl (Anseriformes). This fact was once used to reinforce the opinion of some avian systematists that flamingos and ducks are closely related (Hopkins 1942). Modern phylogenetic work, however, shows that ducks are sister to gamefowl (Galliformes), while flamingos are more closely related to grebes (Dyke and van Tuinen 2004; Hackett et al. 2008). Sharing of three genera of lice between flamingos and ducks was assumed to be a host switch from ducks to flamingos, perhaps mediated by lice on floating feathers (see chapter 7) (Clay 1974; Olson and Feduccia 1980; Rózsa 1991). However, molecular phylogenetic work reveals that at least one of these shared genera (*Anaticola*) is closely related to lice on grebes (*Aquanirmus*) (Johnson et al. 2006).

The most parsimonious cophylogenetic reconstruction suggests that the common ancestor of *Anaticola* and *Aquanirmus* cospeciated with the common ancestor of flamingos and grebes, followed by *Anaticola* switching from flamingos to ducks (fig. 10.6), rather than the reverse. This interpretation is further supported by the fact that, although *Anaticola* occurs on dozens of species of waterfowl (Price et al. 2003), it is missing from the earliest diverging lineages, such as screamers and whistling-ducks. Hence, the expansive radiation of *Anaticola* appears to have been triggered by a major host switch.

There and back again: Major switching and back-switching

Another major host switch appears to have preceded diversification of body lice across pigeons and doves (Columbiformes) of the world (fig. 10.7A). Columbiform body lice are phylogenetically nested within the body lice of gamefowl (Galliformes) (Johnson et al. 2011b). Although it remains unclear precisely when the original switch to Columbiformes occurred, most phylogenetic congruence between body lice and Columbiformes involves terminal branches in the host phylogeny (Clayton and Johnson 2003). This pattern suggests that the switch from Galliformes to Columbiformes happened long after the radiation of major lineages of Columbiformes. Since most lineages of Columbiformes have body lice, diversification of the group probably involved some combination of host switching and cospeciation after the initial host switch. However, long after colonizing the Columbiformes, one lineage of columbiform body lice apparently switched back to the Galliformes (fig. 10.7B). This second switch appears to have been restricted to the Australasian megapodes (mound-builders), although more sampling is needed. This switch did not trigger another major radiation of body lice, except perhaps within the megapodes, presumably because of competition from body lice already present on the other major lineages of Galliformes. It would be interesting to test the competition hypothesis experimentally (chapter 6).

Comparative cophylogenetics of ecological replicate lice

A powerful approach for testing the role of ecological factors in codiversification is to compare similar (unrelated) groups of parasites that share the same hosts—for example, the wing and body lice found on pigeons and doves (fig. 1.12). These "ecological replicates" are similar in many features of their ecology, such as life cycle, diet, and ability to establish viable populations on novel hosts once they have been experimentally transferred to them (fig. 5.7). However, wing and body lice differ in other respects, such as microhabitat use (fig. 1.12), behavior used to escape from host defense (fig. 5.5), competitive interactions (figs. 6.5, 6.7), dispersal ecology (figs. 7.5–7.7), and population structure (fig. 8.4). Body lice are also more host-specific

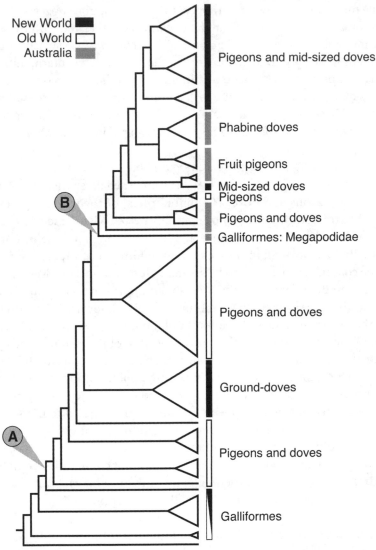

New World ■
Old World □
Australia ▨

Pigeons and mid-sized doves

Phabine doves

Fruit pigeons

■ Mid-sized doves
□ Pigeons

Pigeons and doves

Galliformes: Megapodidae

B

Pigeons and doves

Ground-doves

Pigeons and doves

A

Galliformes

FIGURE 10.7. The phylogeny of body lice from Columbiformes and Galliformes shows that Columbiform lice are nested within Galliform lice, suggesting a host switch from Galliformes to Columbiformes early in the radiation of pigeons and doves (A). Long after colonizing the Columbiformes, one lineage of lice apparently switched back to Galliformes (B). After Johnson et al. (2011b).

tlian wing lice (Johnson et al. 2002b). How, then, does the phylogenetic history of wing and body lice compare to that of their shared hosts? And do the ecological differences between them influence their respective cophylogenetic histories?

The ecological replicate approach works best when both types of parasite can be compared for each species of host included in the analysis. It is also best to compare host-parasite interactions from a single geographic region to prevent broad scale biogeographic patterns from swamping regional patterns. Clayton and Johnson (2003) compared the cophylogenetics of wing and body lice from New World pigeons and doves. Their analysis revealed a striking difference in the cophylogenetic history of the two groups. The phylogeny of body lice (*Physconelloides*) closely mirrors that of the hosts, whereas the phylogeny of wing lice (*Columbicola*) is not correlated with host phylogeny (fig. 10.8). New World body lice show a history of repeated bouts

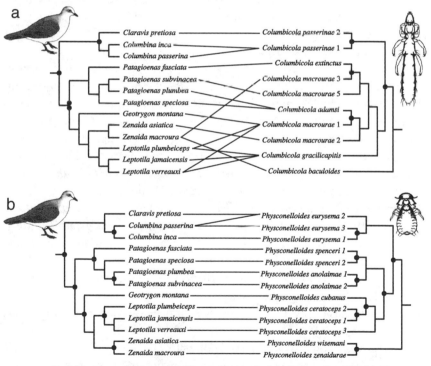

FIGURE 10.8. Cophylogenetic reconstructions of ecological replicates: New World pigeons and doves (left) and their lice (right): (a) *Columbicola* wing lice and (b) *Physconelloides* body lice. The phylogeny for body lice shows extensive congruence with the host phylogeny. In contrast, the phylogeny for wing lice is not more similar to the host phylogeny than expected by chance. Conventions as in fig. 10.2. After Clayton and Johnson (2003).

of cospeciation with their hosts, while wing lice show far less cospeciation. Can ecological data help explain the macroevolutionary events (fig. 9.1) that governed these different cophylogenetic histories? Simply put, why isn't the wing louse phylogeny as congruent as the body louse phylogeny is with the shared host phylogeny?

Sorting events, such as extinction or "missing the boat" (fig. 9.1g,h), are not likely to have played an important role in the evolutionary histories of wing or body lice on pigeons and doves. The prevalence and intensity of both wing and body lice on pigeons and doves tend to be high, particularly in humid regions of the world (fig. 2.5). This, in turn, reduces the likelihood of sorting events. Moreover, both groups of lice have chronic effects on the host (chapter 3), reducing the likelihood of cyclical host-parasite dynamics that might favor sorting events. Most pigeons and doves have moderately widespread geographic distributions, making sorting events less likely. There are exceptions, however, such as species of pigeons and doves that are endemic to oceanic islands (Gibbs et al. 2001). For example, as described in the last chapter, the absence of wing lice from the New Zealand pigeon (*Hemiphaga novaeseelandiae*) may represent a "missing the boat" event when the ancestral host originally colonized New Zealand.

Duplication events (fig. 9.1f) may have played a role in the evolutionary history of both wing and body lice, but they probably do not explain the different cophylogenetic histories. Environmental factors, such as low humidity, create distributional gaps in the wing lice of Columbiformes (Moyer et al. 2002a), similar to those in songbird lice (fig. 10.5). Such gaps may promote duplication of parasites on a single host species. However, one seemingly good example, involving the distribution of *Columbicola macrourae* on mourning doves (*Zenaida macroura*), is not a case of duplication. The drought-tolerant congener *C. baculoides* replaces *C. macrourae* in arid regions of the country (plate 14). On the face of it, this geographic pattern has the appearance of a duplication event. However, the *Columbicola* phylogeny reveals that the two species are descended from distantly related ancestors (plate 13). Moreover, competition between species of *Columbicola* appears to select against the coexistence of different wing lice species on the same species of pigeon or dove; coexistence does occur in some cases, however (chapter 6).

Differences in host switching and cohesion (fig. 9.1b-e, i) are more likely to have contributed to the different cophylogenetic histories of wing and body lice. In many regions of the world, at least some species of pigeons and doves are sympatric and syntopic (sharing habitat) (Gibbs et al. 2001). Overlap between host species creates the potential for host switching and cohesion. These events are governed, in part, by the dispersal ability of lice. Cohesion depends on regular enough dispersal between host species for a single louse species to maintain gene flow. Host switching depends on periodic dispersal to novel host species, followed by establishment of a viable breeding population on the novel host. In chapter 7 we compared the dispersal abilities of wing and body lice. Both groups engage in vertical and direct horizontal transmission; however, only wing lice undergo indirect transmission, through phoretic hitchhiking on hippoboscid flies (plate 5). Even low levels of phoresis can trigger episodes of host switching. The simple fact that wing lice are phoretic, while body lice are not, may explain the dramatically different cophylogenetic patterns shown by these two groups of lice.

Thus, differences in the coevolutionary history of wing and body lice can be explained by differences in host switching that are mediated by members of the broader parasite community (hippoboscid flies) (Harbison and Clayton 2011). Although the contrasting cophylogenetic histories of wing and body lice can be explained by differential phoresis, the question remains as to why wing lice are phoretic in the first place. Phoresis is a risky strategy. Since hippoboscid flies are less specific than wing lice, phoretic lice may often end up on unsuitable hosts and die. Moreover, any benefits incurred in terms of gene flow and avoidance of inbreeding should apply equally to wing lice and body lice. So why are wing lice phoretic, given that body lice are not? The answer may be related to competition between wing and body lice, which is mediated by host defense. We consider this in more detail below. Fig. 10.9 provides a summary of the community ecological interactions of wing and body lice. We now consider how these interactions may influence the micro- and macroevolution of the wing and body lice.

First, preening has a clear selective effect on the body size of lice. Preening also mediates competition between species of lice that share hosts (fig. 6.5). Coadaptation of lice in response to competition can reduce the intensity of the competition. Body lice, which are competitively dominant over wing lice, exert selection on wing lice that presumably reinforces their use of flight feathers as a refuge from body lice, which are never found on flight feathers. Wing lice have effects on body lice, causing them to shift their microhabitat distribution on the body of the host slightly (fig. 6.7d). The two ecomorphs presumably also exert reciprocal selection for efficient harvesting of their common food resource.

Competition-induced microhabitat partitioning may also reinforce host-imposed selection for efficient insertion (wing lice) and burrowing (body lice) behavior on different parts of the host's body (fig. 5.5). The cigar shape of wing lice facilitates insertion between feather barbs, and their long legs facilitate movement over the coarse flight feathers. The long legs are also used to hang onto flies during phoresis (plate 5b).

In short, selection imposed on wing lice by body lice may indirectly re-inforce the phoretic behavior that allows wing lice to disperse to new host individuals, which may be free of body lice. Because hippoboscid flies are

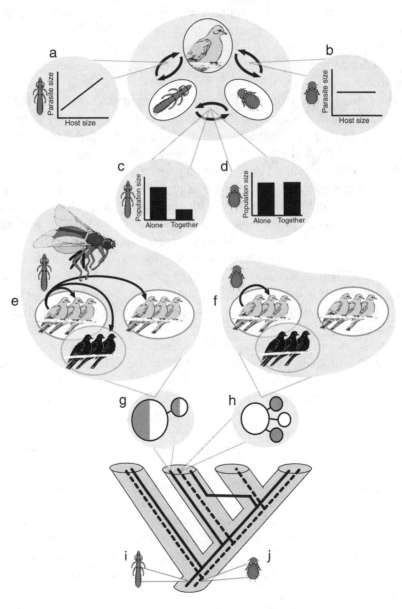

generally not as host specific as most lice, the competition-colonization trade-off presumably increases host switching. Host switching, in turn, erodes cophylogenetic congruence. Thus, the host-mediated competitive interactions between lice may influence patterns of codiversification between birds and lice. In summary, birds, wing lice, body lice, and flies form a complex web of ecological interactions with a clear footprint on cophylogenetic history (fig. 10.9). In the case of wing lice, competition leads to phoresis, which leads to host switching, which leads to an incongruent cophylogenetic pattern, compared to body lice.

In this chapter we have shown how cophylogenetic comparisons, coupled with comparative and experimental data, can be a powerful approach for integrating ecology and history. This integrative approach is particularly useful when applied to ecological replicate systems, such as wing and body lice. In the next chapter, we show how coadaptation between lice and their hosts, and between different species of lice, can trigger coadaptive diversification.

FIGURE 10.9. (*facing page*) Ecological replicate summary: integrating ecology and history. Birds and feather lice exert reciprocal selection for host resistance and parasite escape. Lice exert selection for the evolution of adaptations, such as the bill overhang, that increase the efficiency of preening. Preening, in turn, selects for counter-adaptations that help lice to escape. Wing lice escape from preening by hiding between the barbs of wing feathers that are correlated in size with overall host size. The result is that wing lice follow Harrison's rule of correlated host and parasite body sizes (a). Body lice escape from preening by burrowing into the downy regions of abdominal feathers that are not correlated with overall host size; thus, body lice do not follow Harrison's rule (b). *In summary, coadaptive differences in the phenotypic interface of wing and body lice with the shared host explain differences in body size macroevolution.*

Wing and body lice also compete with one another, with each species altering its distribution on the host in response to the other species. Wing louse populations (c) are smaller when body lice are present. In contrast, body louse populations (d) are unaffected by wing lice. Wing lice can escape from competition through phoretic dispersal; the illustration (e) shows a hippoboscid fly with three wing lice attached (from an actual case). Phoresis enables wing lice to disperse between host individuals, populations, and species. Body lice (f), which do not engage in phoresis, disperse during periods of direct contact between host individuals. Wing lice (g) show less population structure than body lice (h). Wing lice (i) also show less cophylogenetic congruence than body lice (j). *In summary, differences in the interaction of wing and body lice with parasitic flies explain differences in the phylogenetic interface of these lice with shared hosts.*

A major avenue of research in the study of coevolutionary
diversification should be to focus on how and when coevolution
is a source of divergent selection and whether selection on coevolving
traits can be linked to reproductive isolation.
—Althoff et al. 2014

C oadaptive diversification occurs when one of two interacting lineages diversifies in response to coadaptation
between those lineages. The process is also known as
coevolutionary diversification (Althoff et al. 2014). Unfortunately, "coevolution" means different things to
different people (see box 1.1 and chapter 12). Therefore, we prefer "coadaptation," which was the term used by Darwin (see chapter 1). Semantics aside,
Hembry et al. (2014) described coevolutionary diversification as "diversification at any level—within populations and species or across clades—that
is driven by the reciprocal adaptation of interacting species." The authors
point out that the number of rigorous demonstrations of coadaptive diversification can be counted on the fingers of one hand.

In this chapter we review evidence suggesting that bird lice undergo
coadaptive diversification. We begin by considering how coadaptation
between lice and their hosts can trigger the diversification of lice across
host species. We then consider how coadaptation between competing species of lice, mediated by host defense, can contribute to the diversification
of lice within host species.

Coadaptive diversification of lice across host species

Clayton et al. (2003b) used wing lice from doves in Africa, the Galápagos,
and the Philippines to expand the cophylogenetic analysis of New World
Columbicola discussed in the last chapter (fig. 10.8a). This larger data set
revealed statistically significant congruence; however, it also revealed a
good deal of incongruence, partly due to host switching and sharing. Hosts
shared by non-specific *Columbicola* were more similar in body size than
expected by chance, suggesting that size constrains host switching and
sharing (fig. 11.1). Thus, even though wing lice are capable of horizontal
dispersal among host species (chapter 7), they do not appear to be capable
of establishing viable populations on hosts that differ much in size from
the native host.

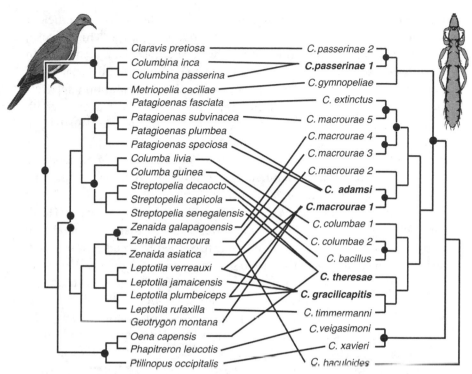

FIGURE 11.1. Cophylogenetic reconstruction of New and Old World pigeons and doves and their *Columbicola* wing lice. Bullets are inferred cospeciation events. The five boldface species of lice are those found on more than one species of host. The hosts shared by these non-specific lice are more similar in body size than expected by chance; 8 of 11 host size differences were within the lowest quartile of all possible differences (Wilcoxon signed rank test, *P* = 0.008). Conventions as in fig. 10.2. After Clayton et al. (2003b).

The influence of body size on host sharing in fig 11.1 echoes the results of the experimental transfers of *C. columbae* in chapter 5 (fig. 5.7.) Those experiments revealed that *C. columbae* could establish viable populations on host species that are much smaller in size than their native rock pigeon host, but only when the hosts were bitted to impair preening (fig. 5.7). In contrast, *C. columbae* could not establish viable populations on small novel hosts capable of normal preening. The reason was that the lice could not fit between the barbs of the smaller flight feathers of small-bodied hosts to escape from preening (fig. 5.8). Thus, preening limits host switching and gene flow among hosts, including hosts to which the lice are capable of dispersing by phoresis or other mechanisms (Clayton et al. 2003b).

Columbicola and other lice exert reciprocal selection on host defense. Specifically, they select for particular sizes of the beak overhang, which is essential for efficient preening (chapter 4). The lice select for efficient host defense, which reciprocally selects for parasite escape. This, in turn, influences coadaptive diversification of the lice because the match between louse body size and host feather size limits the number of host species on which the lice can establish viable populations. This limitation effectively increases the host specificity of the lice, which in turn may favor diversification. *Columbicola* contains 88 described species of wing lice (Bush et al. 2009b), many of which are on sympatric hosts that share habitat (Gibbs et al. 2001). Since only about half of the 300+ known species of pigeons and doves have been carefully examined for lice, it is likely that many additional undescribed species of *Columbicola* exist.

The correlation of *Columbicola* body size with host size, known as Harrison's rule (fig. 5.9a), suggests that soon after lice do manage to establish on a novel host that is somewhat different in size, preening-mediated selection fine-tunes the body size of the lice to maximize their ability to escape from preening (fig. 5.6). Such host switches may be "stepping stones" to the establishment of lice on host species that are even more different from the original native host. Host shifts by herbivorous insects work in a broadly similar fashion (Nyman 2010). Janz (2011) summarizes insect diversification by host switching as follows: "If it [the host] is too similar, colonization is likely but will not lead to disruptive selection; if it is too different, it will not be colonized at all."

To test the hypothesis that preening favors changes in the body size of lice on different-sized hosts, *C. columbae* from wild-caught feral rock pigeons were recently transferred to captive "giant runts"—a domesticated breed of rock pigeon that is larger than feral pigeons. *C. columbae* were also transferred to control feral pigeons. The runts and ferals were first housed in low-humidity animal rooms to kill any lice and eggs already on them (Moyer et al. 2002a). Six months after the lice were transferred (about six louse generations), the body sizes of lice on the giant runts and feral pigeons were compared. The lineages of lice on the two hosts differed significantly in size (fig. 11.2). This result satisfies one criterion for the rigorous demonstration of coadaptive diversification—that is, that coadaptation facilitates divergence between the populations of one of the coadapting species (Althoff et al. 2014). In summary, there is strong evidence to suggest that the coadaptation of host defense and parasite escape leads to divergence in the body size of wing lice on different-sized hosts.

A second criterion for rigorous demonstrations of coevolutionary diversification is that coadapting traits influence reproductive isolation of diverging lineages (Althoff et al. 2014). *C. columbae* wing lice are sexually dimorphic in size. Female metathorax width (a standard measure of size)

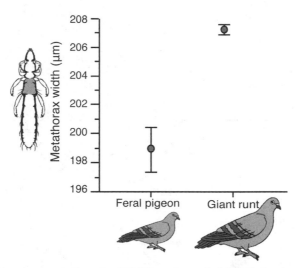

FIGURE 11.2. Mean ± se metathoracic width (shown in grey) of female *C. columbae* wing lice on (control) feral pigeons and "giant runt" pigeons six months following the experimental transfer of lice from a single population originating on feral pigeons (ANOVA, df = 98, P < 0.001). Giant runts are about three times the overall size of feral pigeons. Giant runts have interbarb spaces that are on average 17.5% wider than the interbarb spaces of feral pigeons (Scott Villa et al., unpublished data). Giant runts and feral pigeons preen on a regular basis, and the wing lice on both hosts escape from preening by hiding in the interbarb spaces of the feathers.

is about 5% greater than that of males (fig. 11.3a). This size dimorphism facilitates normal mating, in which a male positions himself beneath the female, grasps her thorax with modified antennae, then curls the tip of his abdomen dorsally to meet hers. Behavioral observations reveal that experimentally paired lice departing much from this level of dimorphism have difficulty remaining in copula, particularly when harassed by competing males trying to dislodge the copulating male.

To test the hypothesis that diverging body sizes of lice influence reproductive isolation, *C. columbae* nymphs were removed from feral pigeons and giant runts and isolated in glass vials with pigeon body feathers (for food). The vials were kept in an incubator at temperature and humidity settings simulating conditions on the host (Malenke et al. 2011). After the lice reached the adult stage, they were photographed and measured under a dissecting microscope. These virgin lice were then randomly paired in new vials, with one male and one female to a vial. After the female louse ultimately died, the vial and the feathers it contained were carefully searched for eggs.

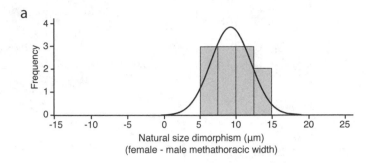

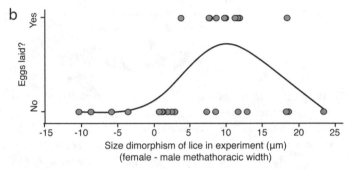

FIGURE 11.3. Influence of sexual dimorphism on egg laying by *Columbicola columbae*, in which females have metathoracic widths averaging 9 μm larger than males (Scott Villa et al., unpublished data). a) Size dimorphism of 162 *C. columbae* collected from 11 wild-caught feral pigeons; b) Significant quadratic relationship between louse reproductive success (egg laying) and size dimorphism (Quadratic Ordinal Logistic Fit, df = 2, Chi Square = 13.1, P < 0.0014). Only females that were 2–18 μm larger than males laid eggs; females exceeding this range of dimorphism in either direction laid no eggs. The model estimates that reproduction is maximized when females are 10.8 μm larger than males.

The results showed that female *C. columbae* laid eggs only if their metathorax width was 2–18 μm larger than the male with which they had been randomly paired (fig. 11.3b). Females exceeding this range of variation, in either direction, laid no eggs. Interestingly, the range of experimental dimorphism that resulted in egg production (fig. 11.3b) was very similar to that of *C. columbae* observed under natural conditions (fig. 11.3a). These data show that the relative body sizes of male and female *C. columbae* influence their reproductive success. Lice that are not well "matched" suffer a dramatic reduction in fitness.

The third criterion to test for rigorous demonstrations of coadaptive diversification is whether coadapting lineages have higher net diversification rates than non-coadapting lineages (Althoff et al. 2014). This hypothesis has not yet been tested broadly for lice, but intriguing patterns involving the relative diversification of head and wing lice are emerging. Hosts control head lice by scratching, and they control wing lice by preening (chapter 4). There is no evidence to suggest that scratching selects for correlated parasite-host body size. In contrast, preening exerts strong selection for correlated wing louse and host body sizes (fig. 5.9a). Thus, coadaptation between host defense (scratching or preening) and parasite escape should have a stronger effect on the diversification of wing lice than head lice. Unfortunately, since pigeons and doves do not have head lice, the hypothesis cannot be tested in this system.

Ducks and geese (Anseriformes) do have both head lice (*Anatoecus*) and wing lice (*Anaticola*). The head louse species *Anatoecus dentatus* is genetically uniform across the different duck species it uses as hosts (unpublished data; Grossi et al. 2014). In contrast, wing lice across the same host species represent different species (unpublished data; Johnson et al. 2006). In the future, it will be interesting to conduct larger-scale comparative studies to continue testing this third criterion of coadaptive diversification using bird lice.

To summarize the results of the work described above, the differential selective effects of preening on the body size of wing lice on different-sized hosts, and the resulting changes in the size of those lice, can lead to reproductive isolation, increasing rates of diversification. Wing lice therefore provide another example of coadaptive diversification.

Coadaptive radiation of lice within host species

Bird lice also show evidence for coadaptive diversification within single host species. This process, which we will call coadaptive radiation, is analogous to the adaptive radiation of free-living species (box 11.1). Feather lice (Ischnocera) exhibit four ecomorphs that vary in morphological and behavioral traits associated with different microhabitats on the body of the host. The four ecomorphs are "head lice," "wing lice," "body lice," and "generalist lice" (fig. 11.4; Johnson et al. 2012). Most orders of birds are host to two or three of these ecomorphs. Some orders, such as Neotropical tinamous, host all four ecomorphs.

Head lice have plump bodies and large, triangular heads (plate 2c; fig. 11.4a). They also have a rostral groove, similar to that of some mammal lice (fig. 5.3a). This groove is thought to help the lice guide feather barbs past the mandibles, with which they grip the barbs to avoid being dislodged by the host when it scratches (chapter 4). Head lice also have expanded temple regions that support large muscles attached to the mandibles, enhancing their grip (Clay 1949b).

BOX 11.1. Adaptive radiation

Where order in variety we see,

And where, though all things differ, all agree.

—Pope 1713

Adaptive radiation occurs when an ancestral species diversifies into descendants that occupy different ecological niches (Dobzhansky 1948; Gavrilets and Losos 2009; Gillespie 2009; Glor 2010; Kassen 2009; Simpson 1953; Wright 1982). Schluter (2000b) summarized the process as follows: "It includes the origin of new species and the evolution of ecological differences between them. It is regarded as the hallmark of adaptive evolution and may well be the most common syndrome in the origin and proliferation of taxa." Although adaptive radiation can be rapid, leading to "bursts" of speciation, not all authors consider speed an essential ingredient. Adaptive radiation has been a topic of widespread investigation, particularly in regions of discontinuous habitat, such as strings of lakes (Meyer et al. 1990), cave systems (Trontelj et al. 2012), and island archipelagos (Gillespie 2004). Classic examples include the adaptive radiation of Darwin's finches on the Galápagos Islands (Grant and Grant 2008) and the silversword plant radiation on the Hawaiian Islands (Seehausen 2004).

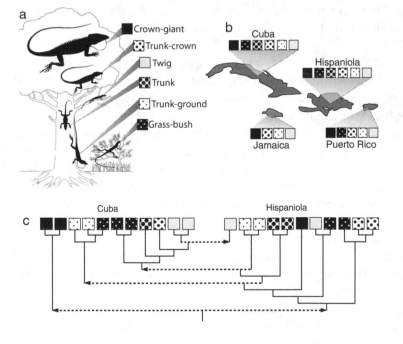

Anolis lizards in the Caribbean are a classic example of adaptive radiation. Species of anoles on a given island have diversified into ecomorphs that differ in size, shape and behavior (a). These ecomorphs live in different microhabitats, such as the ground (grass and bushes), tree-trunks, or tree-crowns (Losos 2009). Because they are descended from a common ancestor, the ecomorphs on a given island are usually more closely related to one another than they are to more similar-looking ecomorphs on other islands (b). Interspecific competition often plays a central role in adaptive radiation. In the case of *Anolis*, sympatric species compete for food and exhibit interspecific aggression and microhabitat shifts in the presence of other species (Losos 2009). Anoles on islands that do not have other species present tend to be generalists that use a greater diversity of microhabitats. The pattern of repeated diversification of *Anolis* ecomorphs among islands is consistent with periodic episodes of dispersal, followed by coadaptation and diversification in response to competition (c, arrows are inferred colonization events; after Losos et al. 1998; Ricklefs and Bermingham 2008; Losos 2009).

Hosts are analogous to islands. Their distributions are discontinuous, creating dispersal barriers for parasites (Janzen 1968; Kuris et al. 1980; Lapoint and Whiteman 2012; Koop et al. 2014). However, these barriers are not absolute; even permanent parasites are capable of periodic dispersal to novel host species. Dispersal of a parasite to a novel host species is analogous to the dispersal of free-living organisms to a new island. If a dispersing parasite establishes a viable breeding population on the novel host species, then a successful host switch will have taken place. In some cases, host switches trigger an adaptive radiation of parasites into different niches on a new host. Adaptive radiations may be responsible for a significant fraction of the impressive diversity of parasites on earth (fig. 1.11; de Meeûs and Renaud 2002; Goater et al. 2013). Unfortunately, much less is known about the process of adaptive radiation in parasites than in free-living organisms.

Wing lice have cigar-like bodies with bullet-shaped heads, and long legs (fig. 11.4b). They spend most of their time on the large flight feathers of the host's wings or tail (Clayton 1991a), where they insert themselves between feather barbs to avoid preening (fig. 5.5a-c). Body lice (fig. 11.4c) have oval bodies and round heads. They live in the abdominal contour feathers, where they avoid preening by dropping between adjacent feathers, or by burrowing into the downy portions of feathers (fig. 5.5d). Generalist lice (fig. 11.4d) have intermediate body shapes. They can be found on most regions of the host's body, where they escape from preening by running quickly.

FIGURE 11.4. Feather louse ecomorphs. (a) Head louse: *Saemundssonia* sp. from a southern fulmar (*Fulmarus glacialoides*); (b) Wing louse: *Columbicola columbae* from a rock pigeon (*Columba livia*); (c) Body louse: *Psittoecus sp.* from a little corella cockatoo (*Cacatua sanguinea*); (d) Generalist louse: *Degeeriella sp.* from a common buzzard (*Buteo buteo*). a, c, and d by V. S. Smith; b after Johnson and Clayton (2003b).

Among host "islands," such as different orders of birds, feather lice have evolved convergent ecomorphs. For example, head lice occur on eight of the ten orders of birds in fig. 11.5. These head lice belong to eight genera that are widely separated on the feather louse phylogeny (plate 16), demonstrating repeated evolution of this ecomorph. The same pattern holds for wing lice, body lice, and generalist lice; each has evolved multiple times (fig. 11.5). These patterns are consistent with repeated adaptive radiations of lice on different host orders, following colonization events by ancestral lice.

The feather louse phylogeny also reveals cases in which sister taxa of lice on the same host group have evolved into different ecomorphs (plate 16; brackets). For example, the genera *Osculotes* and *Pessoaiella*, from hoatzins (Opisthocomidae), are body and generalist lice that are each other's closest relatives. The genera *Pseudolipeurus* and *Pseudophilopterus*, from tinamous (Tinamiformes), are wing lice and head lice that are also sister taxa. Each of these cases, and others on the feather louse phylogeny, may be the result of duplication events, in which the common ancestor speciates on a single host lineage (fig. 9.1f).

Duplication can occur by ecological speciation, in which reproductive isolation evolves in response to divergent natural selection, rather than geographic isolation (Schluter 2000b; Sobel et al. 2010; Guilhem et al. 2012; Nosil 2012). Divergent selection might be generated by differences in host

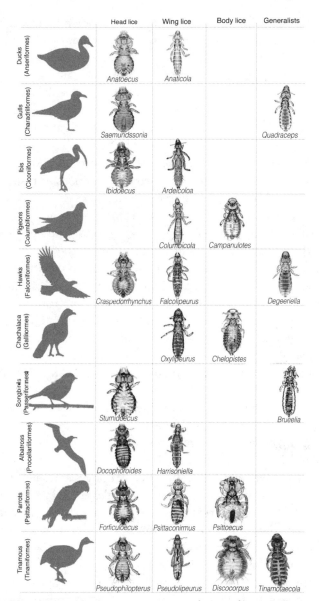

FIGURE 11.5. Distribution of feather louse ecomorphs (columns) across ten orders of birds (rows). Not all orders have all four ecomorphs, but some of the squares without lice are cases in which a specimen was simply not available to photograph. For the phylogenetic relationships of the genera shown here see plate 16. After Johnson et al. (2012).

defense among the different body regions. Birds preen their wings and abdomen, but rely on scratching to control lice on their heads, as discussed earlier in this chapter. Populations of lice living on the head experience different selection pressures than lice living on the wings and body. Moreover, populations of lice living on the wings experience different selection pressures than lice living on the body. This is because preening interacts with large flight feathers differently than it does with abdominal contour feathers.

Why would lice use different microhabitats on the body of the host? One possibility is that microhabitat partitioning arises in response to intraspecific competition for preening-free space in the ancestral louse. Use of different microhabitats may generate rapid divergence in oviposition sites because preening damages louse eggs, not just hatched lice (Nelson and Murray 1971). Lice on the wings attach eggs to the underwing covert feathers to protect them from preening, while lice on the abdomen attach eggs to the downy regions of abdominal feathers. Lice on the head attach eggs to the gular (throat) feathers, which are protected from both preening and scratching (Nelson and Murray 1971; Marshall 1981a). If lice mate in the microhabitat where they lay eggs, which is likely, it could lead to ecological speciation (Nosil 2012).

The generalist ecomorph (fig. 11.4d) may represent the ancestral form of louse, with the other three ecomorphs evolving from it. It may be possible to reconstruct the evolutionary progression of these other ecomorphs. Given the lock-and-key relationship between wing lice and the interbarb space (fig. 5.5.b), evolutionary transitions from wing lice to head lice may be more likely than transitions from head lice to wing lice. Unfortunately, the phylogenetic database for feather lice is not yet sufficient to test this hypothesis. Existing phylogenies show frequent changes in ecomorphology, making it difficult to reconstruct ancestral character states with confidence. With additional phylogenetic data, however, it may be possible to evaluate the direction of ecomorphological change for different radiations of feather lice among major groups of birds.

Competition is not an absolute requirement for adaptive radiation. A lineage can diversify as long as there is ecological opportunity, such as when a novel lineage colonizes a remote archipelago (Schluter 2000b). If a major host switch allows a parasite to colonize a new order of hosts, as illustrated in chapter 10, it may lead to a burst of speciation as the parasite radiates across the different members of that order. Periodic host switching, coupled with host-imposed selection, leads to the evolution of correlated host and parasite body size (Harrison's rule; chapter 5), even without interspecific competition playing any role.

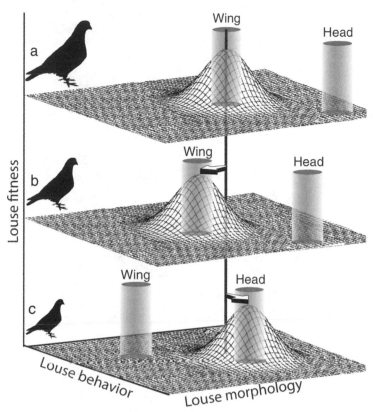

FIGURE 11.6. Adaptive landscapes showing the relative fitness of wing lice in two microhabitats (columns) on different-sized hosts. (a) Combination of behavior (e.g., microhabitat preference) and morphology (e.g., size) that maximizes the fitness of wing lice (*Columbicola*) on their native rock pigeon host. (b) On a slightly smaller novel host, selection favors a decrease in size with no change in microhabitat use. (c) Conversely, on an even smaller novel host, selection favors a shift in microhabitat preference.

Periodic host switching may also trigger transitions between ecomorphs of feather lice. Whether adaptive radiation leads primarily to quantitative changes in parasites, such as altered size, or more qualitative changes, such as different ecomorphs, may depend on the level of difference between the native and novel host (fig. 11.6a). When novel hosts are similar to the native host, adaptive radiation may take the form of slight adjustments in size (fig. 11.6b). However, when novel hosts are extremely different from the native host, the path of least resistance may be for one ecomorph to evolve into another ecomorph (fig. 11.6c).

Behavioral plasticity may be fundamental to this process. Mayr (1963) argued that "a shift into a new niche or adaptive zone is almost without

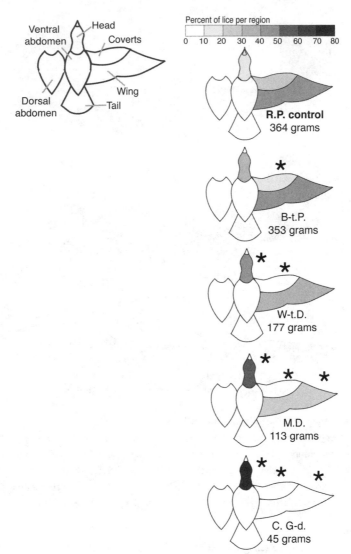

FIGURE 11.7. Microhabitat distributions of rock pigeon wing lice (*Columbicola columbae*) transferred to different species of captive pigeons and doves. As body size decreases (top to bottom of the page) lice shift away from the wings onto the head (*P < 0.05, post-ANOVA Fisher PLSD tests by region). R.P. = rock pigeon, B-t.P. = band-tailed pigeon, W-t.D. = white-tipped dove, M.D. = mourning dove, C.G-d. = common ground dove. After Bush (2009). For scientific names see figure 5.7.

exception initiated by a change in behavior." Baldwin (1896) proposed that behavioral flexibility allows species to persist in novel environments, pending local adaptation. This "Baldwin effect" (Crispo 2007; Scheiner 2104) is thought to play an important role in the ability of herbivorous insects to switch hosts (Agosta et al. 2010; Agosta and Klemens 2008). It has also been documented for free-living organisms in novel habitats, such as dark-eyed juncos (*Junco hyemalis*) moving from temperate montane regions into coastal areas with a more Mediterranean climate (Yeh and Price 2004).

Bush (2009) tested whether behavioral plasticity facilitates host switching in feather lice. She transferred rock pigeon wing lice (*Columbicola columbae*) to a range of smaller host species and, once the lice had established populations for two generations, she compared microhabitat preferences of the lice on the different-sized hosts. Lice shifted their microhabitats on novel hosts, with the magnitude of the shift depending on host size (fig. 11.7). On common ground doves, the smallest host, 100% of wing lice were found on the head! Ultimately, wing lice living on the heads of smaller hosts might experience the Baldwin effect—that is, their behavioral plasticity may "buy time" until morphology better suited to the head evolves. In other words, behavioral plasticity may facilitate the evolution of wing lice into head lice (fig. 11.6). It would be interesting to test this hypothesis using an experimental evolution approach.

In summary, lineages of feather lice on different host species live largely in isolation, punctuated by occasional dispersal between hosts. Isolation causes lice to diverge in behavior and morphology in response to interactions with their respective host species, yielding the coadaptive diversification of lice across host species. Lice also compete with other species of lice, further contributing to the coadaptive radiation of lice. Divergence and speciation, followed by periodic dispersal, may also result in divergent lineages coming into secondary contact with former conspecifics, triggering even more competition and diversification. In short, the distribution of feather louse ecomorphs across the feather louse phylogeny is a consequence of infrequent dispersal events, followed by the repeated, independent evolution of similar ecomorphs.

V SYNTHESIS

12 COEVOLUTION OF LIFE ON HOSTS

A beaver ear stuffed with mites bears a decided resemblance
to a cherry inflorescence stuffed with aphids.
—Janzen 1985b

The overriding goal of this book, as stated in the preface, is to address this question: "how do ecological interactions influence patterns of codiversification?" To this end, we have taken a broad conceptual approach to coevolution and applied it to host-parasite systems, particularly those involving parasitic lice. Chapter 1 provided an introduction to coevolution, followed by an overview of the biology of lice in chapters 2–3. In chapters 4–6, we reviewed the evidence for coadaptation between lice and their hosts, and between competing species of lice on the same host. In chapters 7–8, we reviewed the dispersal and population structure of lice and other parasites. In chapters 9–11, we illustrated how ecological factors can influence the macroevolutionary events that govern host-parasite codiversification.

Janz (2011) suggested using the term coevolution as a broad framework that "help[s] to shift the focus from the potentially pointless discussion of whether a specific pattern qualifies as 'real' coevolution to the much more interesting question of what particular (coevolutionary) processes have shaped the interaction under study and of their possible macroevolutionary effects." We agree with this stance. In contrast, restricting coevolution to coadaptation (box 1.1) is the logical equivalent of restricting evolution to adaptation. The broader definition of coevolution emphasizes the importance of integrating ecology and history.

Throughout this book we have used "coadaptation" to designate coevolution in response to reciprocal selection in microevolutionary (ecological) time. Similarly, we have used "codiversification" to indicate correlated diversification of interacting lineages over macroevolutionary (historical) time. Cospeciation is the subset of cases in which the lineages codiverge simultaneously. Codiversification, which is a broader concept, includes cases in which lineages show correlated patterns of diversification overall, even though they do not diversify at the same time.

In this chapter we conclude the book by introducing a graphical framework that integrates the coadaptation and codiversification "ends" of co-

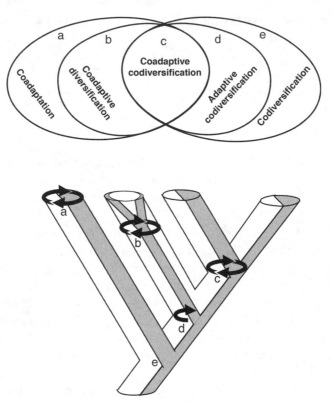

FIGURE 12.1. Coevolutionary framework integrating coadaptation and codiversification. Letters shared by the upper Venn diagram and lower cophylogenetic tree indicate five zones of coevolution: (a) Coadaptation = joint evolution in response to reciprocal selection (black arrows); (b) Coadaptive diversification = divergence of one lineage in response to reciprocal selection (black arrows); (c) Coadaptive codiversification = codivergence of two lineages in response to reciprocal selection (black arrows); (d) Adaptive codiversification = codivergence of two lineages in response to unidirectional selection (black arrow) on one of them; (e) Codiversification = codivergence of two lineages without any selection.

evolution. This framework (fig. 12.1) can be used with any host-parasite system. Indeed, it can be used with any coevolving system. Differences in the five zones of coevolution (12.1a-e) are subtle, yet operational.

Delineating the different zones of coevolution helps to clarify the ways in which dispersal and selection influence the adaptation, coadaptation, diversification, and codiversification of interacting groups. Coevolution, as we use it, varies from a purely microevolutionary focus in the case of coadaptation (fig 12.1a), to a purely macroevolutionary focus in the case of codiversification (fig 12.1e). We discuss these extremes in more detail below.

We then consider the three combinations of adaptation and diversification that make up the middle portions of the framework (fig. 12.1b-d).

When is it coadaptation?

Coadaptation, which is the joint evolution of interacting species in response to reciprocal selection in microevolutionary time, is represented by fig. 12.1a. Coadaptation creates genetic correlations between the non-mixing gene pools of interacting species (Thompson 1982, 1994, 2005). These genetic correlations have a fundamental influence on the evolutionary possibilities of interacting species, both in micro- and macroevolutionary time (Wade 2007; Tellier et al. 2010). Coadaptation is widespread. For example, some form of host-parasite coadaptation has played a role in the evolution of most, if not all, animals, plants, and microbes (Thompson 2005; Brown and Tellier 2011; Schmid-Hempel 2011).

However, coadaptation is difficult to document, particularly under natural conditions. Measuring unidirectional evolution in response to a given selective agent is challenging (Kingsolver et al. 2001). It should therefore come as no surprise that measuring reciprocal selection is also challenging. The intimacy of host-parasite interactions does not necessarily improve their tractability. Hosts have extremely diverse communities of parasites (fig. 1.11). Teasing apart the reciprocal selective effects between a host species and each member of its parasite community is difficult.

Reciprocally, few parasites complete their life cycle in a single species of host. Many parasites have life cycles requiring hosts that belong to different phyla. Moreover, few parasites are truly host specific. Some parasites can use a variety of host species at each stage in their life cycle. Thus, although hosts and parasites are often in close ecological contact for much, if not most, of their lives, hosts are often exposed to the selective effects of several parasite species. Similarly, parasites are exposed to the selective effects of several host species. Parasites can also be exposed to the selective effects of other parasite species on the same hosts.

These complexities notwithstanding, host-parasite coadaptation is conceptually a straightforward process. If the negative effect of a parasite on host fitness is correlated with phenotypic variation in host defense, then the parasite will select for better defense. If the phenotypic variation in host defense has a heritable component, then selection by the parasite will lead to the evolution of improved defense. This scenario assumes no antagonistic pleiotropy or other multi-locus effects that prevent a response to selection. Reciprocally, if improved host defense has a negative effect on parasite fitness, and if that negative effect is correlated with variation in the ability of the parasite to avoid host defense, then the host will select for improved parasite escape. If phenotypic variation in escape has a heritable component, then selection by the host should drive the evolution of improved parasite escape.

Guidelines for studying coadaptation can be found in Gomulkiewicz et al. (2007), Nuismer et al. (2010), and Brockhurst and Koskella (2013). Thompson (2013) reviews recent studies of coadaptation, ranging from field studies of geographic mosaics, to the experimental coevolution of laboratory models.

When is it codiversification?

Codiversification is the correlated diversification of interacting lineages due, at least in part, to their ecological interaction. Codiversification need not be simultaneous, although it is, in fact, simultaneous in the case of cospeciation. Documenting codiversification requires study at much deeper timescales than coadaptation. Unlike coadaptation, which occurs in microevolutionary (ecological) time, codiversification occurs in macroevolutionary time and thus requires cophylogenetic data to document. Fig. 12.1e represents *simple* codiversification, which is the correlated divergence of interacting lineages without any selection. Unlike coadaptation, which is driven by reciprocal selection, simple codiversification is selectively neutral. It is sometimes called "phylogenetic tracking" (Thompson 1994; Althoff et al. 2014), or "phyletic tracking" (Clayton et al. 2003a). However, because these alternatives are used in different ways by different authors (cf. Brooks and McLennan 1991, 2002), we have shied away from using them.

Simple codiversification (no selection) is uncommon. Population genetics theory tells us that, in the absence of selection, it takes but a trickle of gene flow to prevent populations from diversifying (or codiversifying) (Sobel et al. 2010). Thus, simple codiversification will occur only when there is exceedingly low dispersal and virtually no gene flow between populations. Codiversification does occur in some systems where the members of the interacting groups are very closely associated. This is true when ecological circumstances favor the process of cospeciation. For example, simple codiversification occurs between some endosymbiotic bacteria and their hosts (Clark et al. 2000; Takiya et al. 2006), but see box 2.1 for a counter example. It can also occur between permanent parasites and their hosts. Pocket gophers and their lice (fig. 10.1) are a likely example of simple codiversification, which takes place when parasites are more or less stranded on diverging host islands.

Even in the case of very close associates, however, it is difficult to exclude the possibility that selection has played a role in codiversification. In the case of the permanent feather lice of pigeons and doves, for example, host defense (preening) reinforces codiversification by preventing gene flow between parasite populations on different host species (Clayton et al. 2003b). Once selection is involved, codiversification merges into adaptive codiversification, which we consider next.

Adaptive codiversification

Fig. 12.1d represents adaptive codiversification, which is the correlated divergence of interacting lineages in response to unidirectional selection on one of the lineages. It is not possible to separate this process from simple codiversification without ecological data (ideally, experimental data). Although cophylogenetic studies are essential for documenting patterns of codiversification, they are not sufficient to test the underlying mechanisms responsible for it. It is necessary to integrate cophylogenetic data with controlled experiments or other sources of physiological or ecological data (Brooks and McLennan 2002). The relative fitness of parasites on different host lineages can be assessed using controlled transfer experiments, as illustrated in chapter 5.

Experiments designed to test factors influencing host use are of importance in understanding the adaptive codiversification of life on hosts. Host switching is a topic of particular importance. The ecology of host switching is a complex topic with system-specific properties (Janz 2011). Furthermore the cophylogenetic consequences of host switching depend on other events associated with it, such as speciation or extinction (fig. 9.1b-e). As we showed in chapter 10, within a single closely related group of parasites (bird lice), host switching can range from moderately common (fig 10.2), to very common (fig. 10.4), to major switching and back-switching between orders of birds (figs. 10.6, 10.7).

How, then, is it possible to reconstruct the influence of host switching on cophylogenetic history? One approach is to compare ecological replicates—for example, phylogenetically independent parasites that are ecologically similar and share hosts. When the same analytical approach is used for the two parasite groups, differences at the phylogenetic interface are likely to be real (Johnson and Clayton 2003b). One can then explore ecological differences between the two groups that may explain differences in their cophylogenetic histories. In some cases it may even be possible to design experiments or comparative analyses to test hypotheses concerning the influence of specific ecological factors on the phylogenetic interface. Throughout this book (chapters 1, 2, 6, 8, 10, and 11) we have illustrated the ecological replicate approach using the wing and body lice of pigeons and doves. Figure 10.9 summarizes the results of that work.

The ecological replicate approach can be applied to other systems. For example, tropical figs have mutualistic and parasitic fig wasps that share host species (fig. 12.2). Mutualist fig wasps carrying pollen push their way into the internal flower chambers of figs and lay their eggs on the flowers. Upon hatching, the larvae eat some of the developing fig seeds. The wasps mate in the fig cavity after they mature. The fig plant incurs a net benefit from being pollinated by the female wasps as they move over the flowers laying their eggs. In contrast, parasitic fig wasps do not enter or pollinate flowers, but pierce the outside of a fig's wall with a long ovipositor. They deposit eggs in the fig and induce the formation of galls in the fig's ovaries. Unlike mutualist wasps, parasitic wasps provide no benefit to the fig.

Weiblen and Bush (2002) found that mutualistic wasps cospeciate more than parasitic wasps with the figs (fig. 12.2). Parasitic wasps appear to engage in more host switching and sympatric speciation (duplication) than mutualistic wasps. The reason for these differences could be that the reproductive schedules of mutualist wasps and figs are more tightly linked than are those of parasitic wasps and figs.

The use of ecological replicates is essentially a simple comparative approach. More sophisticated comparative analyses can also be used to study host switching (e.g., plate 13) and other topics related to adaptive codiversification and the other zones of coevolution. For example, as we showed in chapter 10, major host switching events have been identified through the cophylogenetic analysis of birds and their lice. Over the louse phylogeny, character reconstruction techniques could also be used to map out changes in microhabitat specialization. The question as to whether changes in ecomorphs are associated with major host switching events might be addressed in this way.

Generally speaking, there is a great deal of untapped potential for studies that integrate cophylogenetic approaches and comparative methods to gain a more comprehensive understanding of adaptive codiversification and the other zones of coevolution.

In the next section, we examine the flip side of the coevolutionary coin. Instead of focusing on the role of unidirectional selection and adaptation in codiversification, we consider the influence of coadaptation on diversification.

Coadaptive diversification

Fig. 12.1b illustrates coadaptive diversification, which occurs when coadaptation triggers the divergence of one of two interacting lineages. This process is similar to coevolutionary diversification, or diversifying coevolution (box 1.1). Coadaptation can also lead to the divergence of *both* interacting

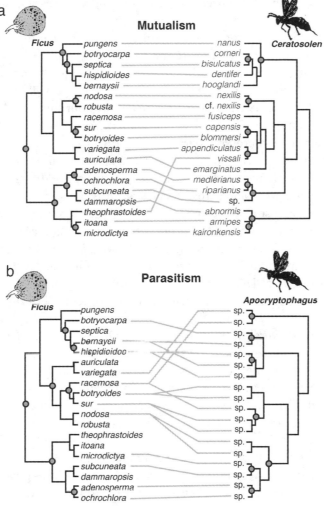

FIGURE 12.2. Cophylogenetic comparisons of *Ficus* figs and their (a) mutalist fig wasps (*Ceratosolen*), and (b) parasitic fig wasps (*Apocryptophagus*). Grey lines connect figs with their associated wasps. The phylogenies of mutualist wasps and figs show a good deal of congruence, indicating that cospeciation has been pervasive between these groups. The phylogenies of parasitic wasps and figs are less congruent, suggesting that host switching and duplication have been more common between these groups. Bullets on nodes are inferred cospeciation events. After Weiblen and Bush (2002).

lineages (fig. 12.1c); this process of coadaptive codiversification is explored in the next section.

One of the most rigorous case studies of coadaptive diversification involves work on red crossbills (*Loxia curvirostra*) and Rocky Mountain lodgepole pines (*Pinus contorta*) (see chapter 1). Seed predators (squirrels) mediate coadaptation between the crossbills and pines, facilitating diversification of crossbills across regions with and without squirrels (Benkman 1999, 2003; Benkman et al. 2001, 2010). Coadaptive diversification has also been shown for red crossbills and black spruce (*Picea mariana*) in New-foundland (Parchman and Benkman 2002; Benkman et al. 2010).

In chapter 11 we reviewed evidence for the coadaptive diversification of bird lice, both within and across species of birds. Birds and their lice are one of the few host-parasite systems in which coadaptive diversification has been rigorously tested. The process may be common across the rich diversity of animal host-parasite systems. However, tests of coadaptive diversification are difficult because they require the integration of ecological and historical data. This, in turn, requires robust phylogenies for both of the interacting groups, and well-designed experiments performed in ecological time.

Coadaptive diversification can also occur in response to competition (Schluter 2010). In chapter 6 we reviewed evidence that competition-colonization trade-offs influence patterns of host-parasite codiversification. We showed that abiotic conditions, such as ambient humidity, can mediate condition-dependent competition between parasites and that this, in turn, influences patterns of diversification. We reviewed evidence that parasite competition, mediated by host defense, governs the coadaptive radiation of parasites on hosts, providing another example of coadaptive diversification.

Experimental coevolution, introduced in chapter 1, holds great promise for research on coadaptive diversification (Brockhurst and Koskella 2013). Experiments involving microbial interactions, such as bacteria and their viruses, can run for hundreds or thousands of microbial generations. Most experimental coevolution studies are, by necessity, lab based (but see Gó-mez and Buckling 2011). Moreover, because they require fast-reproducing organisms, experimental coevolution studies are largely restricted to model organisms, such as microbes, snails, worms, and insects.

Note that, despite the analytical power of experimental coevolution, Buckling et al. (2009) cautioned that such studies are not truly macroevolutionary because gene diversification is not equivalent to species diversification: "Even experiments involving thousands of generations might not be sufficient to understand the underlying causes of some of the differences in . . . species that have diverged from each other over millions of years."

In summary, although experimental coevolution is a powerful approach for studies of coadaptive diversification, it is not a cure-all. Cophylogenetic approaches are also powerful because they can, in principle, be applied to any group of organisms from any habitat on earth. However, rigorous cophylogenetic analysis requires comparison of both the topology and branch lengths of robust phylogenies (Vienne et al. 2013).

Coadaptive codiversification

Fig. 12.1c shows coadaptive codiversification, in which coadaptation causes the reciprocal divergence of interacting lineages. Coadaptive codiversification may be a major factor in the remarkable diversification of herbivorous insects and their plant hosts (Ehrlich and Raven 1964; Mitter et al. 1988; Jaenike 1990; Farrell 1998; Singer and Stireman 2005; Futuyma and Agrawal 2009; Janz 20011; Coley and Kursar 2014). Ehrlich and Raven (1964) proposed the most influential model of this process, which Thompson (1989) called *escape-and-radiate* coevolution (box 12.1). Although escape-and-radiate has the potential to explain the diversification of both hosts and parasites, it does not predict congruent phylogenies because the coadaptive radiations it anticipates are not simultaneous (Thompson 1989; Futuyma and Agrawal 2009; Segraves 2010; Janz 2011). Phylogenetic evidence consistent with escape-and-radiate coevolution has eluded investigators (Futuyma and Agrawal 2009). However, some of the specific predictions of the model have received support (Segraves 2010).

One prediction is that parasites will track host resources rather than host phylogenies. This process of "resource tracking" (Kethley and Johnston 1975) is a type of ecological fitting (Janzen 1985b). It occurs when the members of one group track the members of another group because they represent a resource needed by the first group. In cases where the distribution of the resource is not congruent with the group's phylogenetic history, resource tracking will lead to phylogenetic incongruence. For example, a cophylogenetic comparison of phytophagous beetles (*Blepharida*) and their host plants (*Bursera*) found that host switching among plant lineages was correlated with host chemical similarity, not host phylogeny (fig. 12.3). It is important to bear in mind, however, that in cases where the host resource happens to be congruent with host phylogenetic history, resource tracking can actually reinforce phylogenetic congruence (Clayton et al. 2003a). Testing for resource tracking provides yet another example of the importance of coupling cophylogenetic and ecological approaches to coevolution.

A second prediction of the escape-and-radiate model is that speciation rates of herbivores will increase following a major host switch. Evidence consistent with this prediction exists for leaf-mining flies, which show

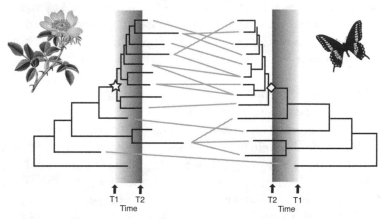

BOX 12.1. Escape-and-radiate coevolution

Ehrlich and Raven (1964) proposed a model of coadaptive evolution and codiversification between butterflies and their food plants. Their "escape-and-radiate" model, illustrated here, is an attempt to explain the incredible diversity of these groups, as well as other herbivorous insects and plants (Janz 2011). Most plants use chemical defenses to combat insects. These defenses select for insects that can circumvent them. At some point in time (T1 Star) a plant lineage evolves a novel defense that cannot be circumvented by the insects. During the ensuing period of "escape," the plant lineage rapidly diversifies as its populations, now free from parasite pressure, expand and radiate into different environmental niches. At a later point in time (T2 Diamond) an insect lineage evolves a counter-defense that allows it to overcome the new plant defense. This new counter-defense allows the insect lineage to expand and radiate onto the recently evolved lineages of plants. The insect radiation follows the plant radiation in time. Therefore, the escape and radiate process consists of alternating bursts of diversification between the plants and insects. Escape-and-radiate coevolution may also apply to some animal host-parasite systems. Diagram modified from Segraves (2010).

pulses of diversification after host shifts onto new plant clades (Winkler et al. 2009). Speciation has also been associated with host switching in pinyon pine aphids (Favret and Voegtlin 2004). Generally speaking, broad phylogenetic comparisons support the notion that insect diversification and plant diversification are related (Farrell and Mitter 1994).

One escape-and-radiate assumption that is particularly difficult to test is the prediction that plant lineages will show an increase in diversification following a *decrease* in herbivore pressure. Unfortunately, it is difficult to

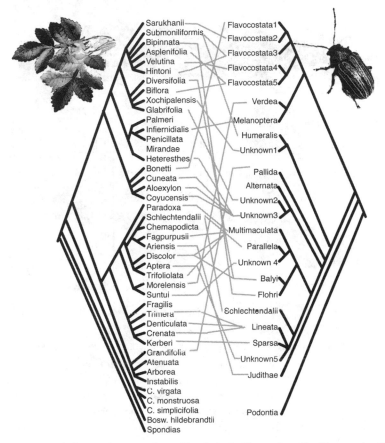

FIGURE 12.3. Cophylogenetic comparison of host plants (*Bursera*) and herbivorous beetles (*Blepharida*). Grey lines connect host plants with their associated beetles. The beetle phylogeny, which is not congruent with the host plant phylogeny, shows evidence of frequent host switching. Plants attacked by related beetles had similar chemistry. After Becerra (1997).

demonstrate that a group of plants was *not* fed upon by a given group of herbivores at some point in the past. Thus, a counterintuitive feature of the escape-and-radiate model is that diversification of plants and insects occurs, not in response to selection, but in response to the *absence* of selection created by the evolution of the new defense or counter defense. Diversification is triggered by ecological release into enemy-free space, or onto novel food resources. The model is still broadly coadaptive because staggered reciprocal selection defines the boundaries of ecological opportunities for diversification (Hembry et al. 2014). As summarized by Thompson (2013), "The pattern of escape-and-radiate coevolution appears at higher

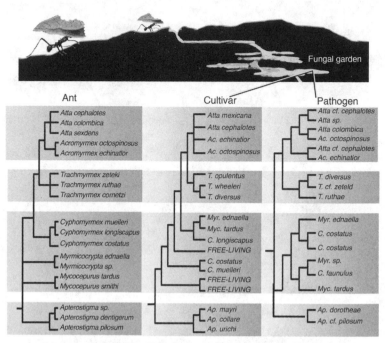

FIGURE 12.4. Tripartite codiversification of ants (left), fungal mutualists (middle), and fungal pathogens (right). Cultivar and pathogen names refer to the ant species with which the fungi are associated (but not all ant names appear in the ant phylogeny). Basal clades are congruent (shaded groups). More recent branches are not congruent. For example, in the case of the ant species *Mycocepurus tardus,* neither the cultivar lineage, nor the pathogen lineage (both indicated by *Myc. tardus*), correspond in phylogenetic position to the ant. After Currie et al. (2003).

taxonomic levels, where each starburst of species on one side of the interaction is matched later with a starburst of speciation on other side."

Other processes of codiversification, such as cospeciation, are similarly scale dependent. For example, attine ants cultivate mutualistic fungal gardens for food. These fungal gardens can be infected by fungal pathogens. Currie et al. (2003) studied tripartite coevolution among these groups by comparing the phylogenies of the ants, fungal cultivars, and fungal pathogens (fig. 12.4). At the deepest phylogenetic level, the three phylogenies were congruent, despite 50 million years of evolution. At more recent phylogenetic levels, however, there is evidence for occasional host switching, which complicates the patterns of codiversification (Gerardo et al. 2006). Note that if one focuses only on the more recent clades, the ancient pattern of cospeciation might be overlooked.

Diversification by host switching is a common feature of host-parasite coevolution (chapters 9 and 10). Studies of herbivorous insects suggest that up to 50% of all speciation events within herbivorous clades are associated with host shifts (Winkler and Mitter 2008; Nyman et al. 2010). Diversification by host switching is also important in animal host-parasite systems (chapter 10; Zietara and Lumme 2002; Tellier et al. 2010; de Vienne et al. 2013). Note, however, that coadaptive codiversification does not depend on diversification by host switching. Host-imposed selection can still have profound effects on the diversification of parasites without any host switching (Schmid-Hempel 2011). For example, host defense can reinforce cospeciation by *preventing* host switching (Clayton et al. 2003b). Mathematical models and empirical studies generally show that host defense contributes to the genetic variation and diversification of parasites (Best et al. 2010; Paterson et al. 2010; Schulte et al. 2013; Brockhurst and Koskella 2013). The more pressing question is how often diversification of parasites has a reciprocal selective effect on the diversification of hosts.

Haldane (1949) was among the first to suggest that parasites frequently have reciprocal effects on the genetic variation and diversification of hosts. Since Haldane's paper, the potential role of parasite-mediated selection in host diversification has received steadily increasing attention (Price 1980; Hamilton and Zuk 1982; Price et al. 1986; Buckling and Rainey 2002; Ricklefs 2010a,b; Loker 2012; Karvonen and Seehausen 2012; Brockhurst and Koskella 2013). Theoretical work confirms that reciprocal coadaptation between parasites and their hosts facilitates host diversification (Duffy and Forde 2009; Yoder and Nuismer 2010). Moreover, recent empirical work demonstrates that experimental coadaptation can drive reproductive isolation of hosts (box 12.2). Additional experimental tests of coadaptive codiversification are needed in natural populations.

Conclusion: Integrating ecology and history

Coadaptation (fig. 12.1a) is the only coevolutionary process that can be tested solely in ecological time. It requires documenting reciprocal selection and joint evolutionary response. In contrast, work on codiversification (fig. 12.1e) requires historical (cophylogenetic) data. Codiversification can take place simply because the members of one group, such as permanent parasites, are "trapped" on diversifying host islands. Realistically, however, cases in which selection plays no role are probably rare. Therefore, careful tests of codiversification also require ecological data. Specifically, they require methods designed to detect selection and local adaptation. For example, the relative fitness of parasites on different host lineages can be assessed using controlled transfer experiments. When evidence of local

BOX 12.2. Parasite-mediated host diversification

Host diversification has profound effects on parasite diversification. But does parasite diversification have reciprocal effects on host diversification? Factors unrelated to parasites, such as geographic barriers, environmental heterogeneity, and other community-level interactions, are central to the diversification of host lineages. However, some attention has also been given to the role that pathogens and other parasites play in the diversification of their hosts. A number of mechanisms have been proposed for how parasite-mediated host diversification can happen (Brucker and Bordenstein 2012; Karvonen and Seehausen 2012).

First, parasites may help reinforce the reproductive isolation of host lineages once they are already diverging (Thompson 1987). This idea is related to the hybrid susceptibility hypothesis, in which hybrids are less resistant to parasites, and have higher parasite loads, than parental species (Fritz et al. 1999). For example, a study of the parasites of mice in a European hybrid zone between *Mus musculus* and *M. domesticus* showed that the abundance of pinworms in hybrid mice was an order of magnitude higher than that in non-hybrid mice (Sage et al. 1986). Genetic and environmental factors were confounded in this study, but a later study with mice in a different region of the same hybrid zone suggested that the high parasite load of hybrids was not due purely to environmental variation (Moulia et al. 1991). A possible mechanism for this phenomenon is that the two species of mice have different alleles for resistance to worms, but recombinant hybrid backcrosses lose the best allelic combinations. Hybridization could lead to negative epistasis between immunity genes from the different species, resulting in resistance that is inferior to that of either parental species (Brucker and Bordenstein 2012).

A more direct mechanism for parasite-mediated host diversification is if parasites select for the origin of reproductive isolation in hosts. Bérénos et al. (2012) allowed five lines of the beetle *Tribolium castaneum* to coevolve for 17 generations with their microsporidian parasite, *Nosema whitei*. This parasite kills beetles when they become infected. Five uninfected control lines of the beetle were also bred over the same period of time. Mortality of the coadapted lines decreased over the course of 17 generations, but mortality of the control lines did not change. To test the effect of the 17 generations of coadaptation on host reproductive isolation, the beetles were crossed both within and between coevolved lines. Beetles in the control lines were also crossed. Crosses between coevolved lines produced fewer offspring than crosses within coevolved lines. There was no difference in the reproductive success of crosses between control lines, compared to crosses within control lines. Moreover, the coevolved lines that differed the most in parasite resistance were the most reproductively isolated. The results of this intriguing study show that host-parasite coadaptation can indeed promote reproductive isolation between host lineages.

Parasites could also contribute to host diversification through indirect mechanisms (Price et al. 1986; Hembry et al. 2014). Host populations sometimes show local (geographic) adaptation in resistance to different parasite strains (Thompson 1987; Duffy and Forde 2009; Yoder and Nuismer 2010). When a host population comes into contact with nonlocal parasites, it may suffer a reduction in fitness that selects against dispersing hosts (MacColl and Chapman 2010). Over time, this process could contribute to the reproductive isolation of host populations.

Diversification of hosts may also be reinforced by parasites mediating the outcome of competition between hosts (Price et al. 1986). In a study of phage viral parasites of bacteria, Buckling and Rainey (2002) showed that parasites increase the diversification of different host populations, while decreasing diversification within host populations because of reduced competition. Parasite-mediated host competition may also prevent the loss of host diversity. Parasites that are disproportionately virulent to competitively superior host species may protect inferior competitors from competitive exclusion (Barbehenn 1969).

Parasites may limit the population sizes of their hosts, which could make host populations more patchily distributed. This, in turn, could lower the geographic range overlap between closely related host species, allowing for increased niche packing and greater host diversification (Ricklefs 2010a). Meanwhile, parasite-free hosts might expand their ranges, allowing for more geographic differentiation and overlap with recently diverged host species (Ricklefs 2010b). Under this scenario, parasites may actually limit host diversification, which could partially explain the frequency of adaptive radiations on islands, which are thought to have fewer parasites (Paterson et al. 1999; Ricklefs 2010b).

In summary, the potential role of parasites in mediating host diversification is ripe for additional study (Karvonen and Seehausen 2012).

adaptation by parasites to particular host lineages is detected, codiversification has at least some adaptive component (12.1d). Note that, in cases of adaptive codiversification, if selection is correlated with host phylogenetic distance, the process may reinforce patterns that are similar to those resulting from simple codiversification (fig. 12.1e).

Coadaptation between interacting taxa can also trigger diversification (fig. 12.1b). This may be true, for example, if divergent selection is linked to reproductive isolation of parasite populations. Another testable prediction is that coadapting lineages have higher net diversification rates than related lineages not engaged in coadaptation. Tests of this prediction are still needed. If divergent selection on parasites leads to reciprocal divergent selection of their hosts (box 12.2), then coadaptive codiversification (fig. 12.1c) may be at work. Escape-and-radiate coevolution, which involves alternating bouts of selection, is the oldest model of this process. Animal hosts and their parasites may also codiversify in response to reciprocal selection. These processes do not require cophylogenetic congruence or cospeciation, and they should not be confused with them.

In summary, distinguishing among the different zones of coevolution (fig 12.1) is even more difficult than remembering their names. The importance of integrating phylogenetic, comparative, and experimental approaches cannot be overstated for studies of coevolution (Weber and Agrawal 2012). It is also essential to understand the natural history of the system (Greene 2005; Thompson 2014). Coevolutionary biology is a challenging field, particularly when one attempts to study it under natural conditions. Is it worth all the effort? Absolutely! The spoils awaiting successful researchers are exciting, as nicely summarized by Futuyma (2010): "Through the eons, spirals of coevolutionary ploy and counterploy, the evolutionary conflicts between enemies and victims, between competitors, even between mutually exploiting symbionts, have produced waste, evolutionary cul-de-sacs, extinction—but also exquisite forms and functions, adaptations no inventor would ever have conceived, and the 'endless forms most beautiful and most wonderful' of which Darwin sang in the last lines of his most wonderful book."

LITERATURE CITED

Able, D. J. 1996. The contagion indicator hypothesis for parasite-mediated sexual selection. *Proceedings of the National Academy of Sciences USA* 93:2229–2233.

Agosta, S. J., and J. A. Klemens. 2008. Ecological fitting by phenotypically flexible genotypes: implications for species associations, community assembly and evolution. *Ecology Letters* 11:1123–1134.

Agosta, S. J., N. Janz, and D. R. Brooks. 2010. How specialists can be generalists: resolving the "parasite paradox" and implications for emerging infectious disease. *Zoologia* 27:151–162.

Agrios, G. N. 2005. *Plant Pathology.* 5th ed. Elsevier Academic Press.

Aguileta, G., G. Refregier, R. Yockteng, E. Fournier, and T. Giraud. 2009. Rapidly evolving genes in pathogens: methods for detecting positive selection and examples among fungi, bacteria, viruses and protists. *Infection Genetics and Evolution* 4:656–670.

Alfaro, J. W. L., L. Matthews, A. H. Boyette, et al. 2011. Anointing variation across wild capuchin populations: a review of material preferences, bout frequency and anointing sociality in Cebus and Sapajus. *American Journal of Primatology* 74:1–16.

Allen, A. 2014. *The Fantastic Laboratory of Dr. Weigl: How Two Brave Scientists Battled Typhus and Sabotaged the Nazis.* W. W. Norton, New York.

Allen, J. M., C. O. Worman, J. E. Light, and D. L. Reed. 2013. Parasitic lice help to fill in the gaps of early hominid history. In *Primates, Pathogens, and Evolution*, edited by J. F. Brinkworth and K. Pechenkina, 161–186. Springer, New York.

Althoff, D. M., K. A. Segraves, and M. T. J. Johnson. 2014. Testing for coevolutionary diversification: linking pattern with process. *Trends in Ecology and Evolution* 29: 82–89.

Alvarez, L. W., W. Alvarez, F. Asaro, and H. V. Michel. 1980. Extraterrestrial cause for the Cretaceous-Tertiary extinction. *Science* 208:1095–1108.

Amarasekare, P. 2003. Competitive coexistence in spatially structured environments: a synthesis. *Ecology Letters* 6:1109–1122.

Amin, O. M., and M. E. Wagner. 1983. Further notes on the function of pronotal combs in fleas (Siphonaptera). *Annals of the Entomological Society of America* 76: 232–234.

Anderson, B., and S. D. Johnson. 2008. The geographical mosaic of coevolution in a plant-pollinator mutualism. *Evolution* 62:220–225.

Anderson, R. M., and R. M. May. 1978. Regulation and stability of host-parasite population interactions. I. Regulatory processes. *Journal of Animal Ecology* 47: 219–247.

Andersson, M. 1994. *Sexual Selection.* Princeton University Press, Princeton.

Ansari, M. 1947. Mallophaga (Ischnocera) infesting birds in the Punjab (India). *Proceedings of the National Institute of Sciences, India* 13:255–303.

Arends, J. J. 1997. External parasites and poultry pests. In *Diseases of Poultry*, 10th ed., edited by B. W. Calnek, 785–813. Iowa State University Press, Ames.

Ash, J. S. 1960. A study of the Mallophaga of birds with particular reference to their ecology. *Ibis* 102:93–110.

Askew, R. R. 1971. *Parasitic Insects.* Heinemann Education Books, London.

Assersohn, C., A. Whiten, Z. T. Kiwede, J. Tinka, and J. Karamagi. 2004. Use of leaves to inspect ectoparasites in wild chimpanzees: a third cultural variant? *Primates* 45:255–258.

Aznar, F. J., M. S. Leonardi, B. Berón-Vera, et al. 2009. Population dynamics of *Antarctophthirus microchir* (Anoplura: Echinophthiriidae) in pups from South American sea lion, *Otaria flavescens*, in Northern Patagonia. *Parasitology* 136: 293–303.

Baker, M. D., C. R. Vossbrinck, J. J. Becnel, and T. G. Andreadis. 1998. Phylogeny of *Amblyospora* (Microsporida: Amblyosporidae) and related genera based on small subunit ribosomal DNA data: a possible example of host parasite cospeciation. *Journal of Invertebrate Pathology* 71:199–206.

Baker, D. A., H. D. Loxdale, and O. R. Edwards. 2003. Genetic variation and founder events in the parasitoid wasp, *Diaeretiella rapae* (M'Intosh) (Hymenoptera: Braconidae: Aphidiidae) affecting its potential as a biological control agent. *Molecular Ecology* 12:3303–3311.

Balakrishnan, C. N., and M. D. Sorenson. 2006. Dispersal ecology versus host specialization as determinants of ectoparasite distribution in brood parasitic indigobirds and their estrildid finch hosts. *Molecular Ecology* 16:217–229.

Baldwin, J. M. 1896. A new factor in evolution. *American Naturalist* 30:441–451.

Banks, J. C., R. L. Palma, and A. M. Paterson. 2006. Cophylogenetic relationships between penguins and their chewing lice. *Journal of Evolutionary Biology* 19: 156–166.

Banks, J. C., and A. M. Paterson. 2005. Multi-host parasite species in cophylogenetic studies. *International Journal for Parasitology* 35:741–746.

Bany, J., A. Pfeffer, and M. D. Phegan. 1995. Comparison of local and systemic responsiveness of lymphocytes *in vitro* to *Bovicola ovis* antigen and concanavalin A in *B. ovis*-infested and naive lambs. *International Journal for Parasitology* 25: 1499–1504.

Barbehenn, K. R. 1969. Host-parasite relationships and species diversity in mammals: an hypothesis. *Biotropica* 1:29–35.

Barker, S. C. 1991. Evolution of host-parasite associations among species of lice and rock-wallabies: coevolution? *International Journal for Parasitology* 21:497–501.

———. 1994. Phylogeny and classification, origins, and evolution of host associations of lice. *International Journal for Parasitology* 24:1285–1291.

Baron, R. W., and J. Weintraub. 1987. Immunological responses to parasitic arthropods. *Parasitology Today* 3:77–82.

Bartlett, C. M. 1993. Lice (Amblycera and Ischnocera) as vectors of Eulimdana spp. (Nematoda: Filarioidea) in Charadriiform birds and the necessity of short reproductive periods in adult worms. *Journal of Parasitology* 79:85–91.

Baum, H. 1968. Biology and ecology of the feather lice of blackbirds [in German]. *Angewandte Parasitologie* 9:129–175.

Becerra, J. X. 1997. Insects on plants: macroevolutionary chemical trends in host use. *Science* 276:253–256.

Bell, J. F., S. J. Stewart, and W. A. Nelson. 1982. Transplant of acquired resistance to *Polyplax serrata* (Phthiraptera: Hoplopleuridae) in skin allografts to athymic mice. *Journal of Medical Entomology* 19:164–168.

Bell, J. F., C. M. Clifford, G. J. Moore, and G. Raymond. 1966. Effects of limb disability on lousiness in mice. III. Gross aspects of acquired resistance. *Experimental Parasitology* 18:49–60.

Bell, J. F., W. L. Jellison, and C. R. Owen. 1962. Effects of limb disability on lousiness in mice. I. Preliminary studies. *Experimental Parasitology* 12:176–183.

Belt, T. 1874. *A Naturalist in Nicaragua*. John Murray, London.

Benjamini, E., B. F. Feingold, and L. Kartman. 1961. Skin reactivity in guinea pigs sensitized to flea bites: the sequence of reactions. *Proceedings of the Society for Experimental Biology and Medicine* 108:700–702.

Benkman, C. W. 1999. The selection mosaic and diversifying coevolution between crossbills and lodgepole pine. *American Naturalist* 153:S75-S91.

———. 2003. Divergent selection drives the adaptive radiation of crossbills. *Evolution* 57:1176–1181.

Benkman, C. W., W. C. Holimon, and J. W. Smith. 2001. The influence of a competitor on the geographic mosaic of coevolution between crossbills and lodgepole pine. *Evolution* 55:282–294.

Benkman, C. W., T. L. Parchman, and E. T. Mezquida. 2010. Patterns of coevolution in the adaptive radiation of crossbills. *Annals of the New York Academy of Sciences* 1206:1–16.

Ben-Yakir, D., K. Y. Mumcuoglu, O. Manor, J. Ochanda, and R. Galun. 2008. Immunization of rabbits with a midgut extract of the human body louse *Pediculus humanus humanus*: the effect of induced resistance on the louse population. *Medical and Veterinary Entomology* 8:114–118.

Berenbaum, M. R. 1991. Coumarins. In *Herbivores: Their Interactions with Secondary Metabolites*, edited by G. A. Rosenthal and M. R. Berenbaum, 221–249. Academic Press, New York.

Berenbaum, M. R., and A. R. Zangerl. 1992. Genetics of physiological and behavioral resistance to host furanocoumerin in the parsnip webworm. *Evolution* 46:1373–1384.

Bérénos, C., P. Schmid-Hempel, and K. M. Wegner. 2012. Antagonistic coevolution accelerates the evolution of reproductive isolation in *Tribolium castaneum*. *American Naturalist* 180:520–528.

Bergstrom, C. T., and L. A. Dugatkin. 2011. *Evolution*. W. W. Norton, New York.

Best, A., A. White, E. Kisdi, J. Antonovics, M. A. Brockhurst, and M. Boots. 2010. The evolution of host-parasite range. *American Naturalist* 176:63–71.

Biard, C., N. Saulnier, M. Gaillard, and J. Moreau. 2010. Carotenoid-based bill colour is an integrative signal of multiple parasite infection in blackbird. *Naturwissenschaften* 97:987–995.

Blanchet, S., O. Rey, and G. Loot. 2010. Evidence for host variation in parasite tolerance in a wild fish population. *Evolutionary Ecology* doi:10.1007/s10682-010-9353-x.

Blanco, G., J. de la Puente, M. Corroto, A. Baz, and J. Colás. 2001. Condition-dependent immune defence in the magpie: how important is ectoparasitism? *Biological Journal of the Linnaean Society* 72:279–286.

Blem, C. R., and L. B. Blem. 1993. Do swallows sunbathe to control ectoparasites? An experimental test. *Condor* 95:728–730.

Blouin, M. S., C. A. Yowell, C. H. Courtney, and J. B. Dame. 1995. Host movement and the genetic structure of parasitic nematodes. *Genetics* 141:1007–1014.

Bolles, R. C. 1960. Grooming behavior in the rat. *Journal of Comparative and Physiological Psychology* 53:306–310.

Bonilla, D. L., L. A. Durden, M. E. Eremeeva, and G. A. Dasch. 2013. The biology and taxonomy of head and body lice: implications for louse-borne disease prevention. *PLoS Pathogens* 9:e1003724.

Bonner, J. T. 2006. *Why Size Matters*. Princeton University Press, Princeton.

Bonser, R. H. C. 1995. Melanin and the abrasion resistance of feathers. *Condor* 97: 590–591.

Booth, D. T., D. H. Clayton, and B. A. Block. 1993. Experimental demonstration of the energetic cost of parasitism in free-ranging hosts. *Proceedings of the Royal Society of London B* 253:125–129.

Borchelt, P. L., and L. Duncan. 1974. Dustbathing and feather lipid in bobwhite (*Colinus virginianus*). *Condor* 76:471–472.

Borgia, G., and K. Collis. 1989. Female choice for parasite-free male satin bowerbirds and the evolution of bright male plumage. *Behavioral Ecology and Sociobiology* 25: 445–453.

Borgia, G., M. Egeth, J. A. Uy, and G. L. Patricelli. 2004. Juvenile infection and male display: testing the bright male hypothesis across individual life histories. *Behavioral Ecology* 15:722–728.

Boulinier, T., K. D. McCoy, and G. Sorci. 2001. Dispersal and parasitism. In *Dispersal*, edited by J. Clobert, E. Danchin, A. A. Dhondt, and J. D. Nichols. Oxford University Press, Oxford.

Boyd, B. M., and D. L. Reed. 2012. Taxonomy of lice and their endosymbiotic bacteria in the post-genomic era. *Clinical Microbiology and Infection* 18:324–331.

Britt, A. G., C. L. Cotton, I. H. Pitman, and A. N. Sinclair. 1986. Effects of the sheep-chewing louse (*Damalinia ovis*) on the epidermis of the Australian Merino. *Australian Journal of Biological Sciences* 39:137–144.

Brockhurst, M. A., and B. Koskella. 2013. Experimental coevolution of species interactions. *Trends in Ecology & Evolution* 28:367–375.

Brodie, E. D., III, and E. D. Brodie Jr. 1999. Predator-prey arms races. *Bioscience* 49: 557–568.

Brodie, E. D., III, and B. J. Ridenhour. 2003. Reciprocal selection at the phenotypic interface of coevolution. *Integrative and Comparative Biology* 43:408–418.

Broennimann, P. 1981. *Auco on the Cononaco*. Birkhäuser, Boston.

Bronstein, J. L. 1994. Conditional outcomes in mutualistic interactions. *Trends in Ecology and Evolution* 9:214–217.

Brooke, M. de L. 1985. The effect of allopreening on tick burdens of molting eudyptid penguins. *Auk* 102:893–895.

———. 2010. Vertical transmission of feather lice between adult blackbirds *Turdus merula* and their nestling: a lousy perspective. *Journal of Parasitology* 96:1076–1080.

Brooke, M. de L., and H. Nakamura. 1998. The acquisition of host-specific feather lice by common cuckoos (*Cuculus canorus*). *Journal of the Zoological Society of London* 244:167–173.

Brooks, D. R. 1979. Testing the context and extent of host-parasite coevolution. *Systematic Biology* 28:299–307.

Brooks, D. R., and D. A. McLennan. 1991. *Phylogeny, Ecology, and Behavior*. University of Chicago Press, Chicago.

———. 2002. *The Nature of Diversity: An Evolutionary Voyage of Discovery*. University of Chicago Press, Chicago.

Brouqui, P. 2011. Arthropod-borne diseases associated with political and social disorder. *Annual Review of Entomology* 56:357–374.

Brown, C. R., M. B. Brown, and B. Rannala. 1995. Ectoparasites reduce long-term survival of their avian host. *Proceedings of the Royal Society of London B* 262:313–319.

Brown, G. K., A. R. Martin, T. K. Roberts, and R. H. Dunstan. 2005. Molecular detection of *Anaplasma platys* in lice collected from dogs in Australia. *Australian Veterinary Journal* 83:101–102.

Brown, J. K. M., and A. Tellier. 2011. Plant-parasite coevolution: bridging the gap between genetics and ecology. *Annual Review of Phytopathology* 49:345–367.

Brown, N. S. 1972. The effect of host beak condition on the size of *Menacanthus stramineus* populations of domestic chickens. *Poultry Science* 51:162–164.

———. 1974. The effect of louse infestation, wet feathers, and relative humidity on the grooming behavior of the domestic chicken. *Poultry Science* 53:1717–1719.

Brown, S. P., R. F. Inglis, and F. Taddei. 2009. Synthesis: evolutionary ecology of microbial wars: within-host competition and (incidental) virulence. *Evolutionary Applications* 2:32–39.

Brucker, R. M., and S. R. Bordenstein. 2012. Speciation by symbiosis. *Trends in Ecology and Evolution* 27:443–451.

Brunetti, O., and H. Cribbs. 1971. California deer deaths due to massive infestation by the louse (*Linognathus africanus*). *California Fish and Game* 57:138–153.

Bruyndonckx, N., S. Dubey, M. Ruedi, and P. Christe. 2009. Molecular cophylogenetic relationships between European bats and their ectoparasitic mites (Acari: Spinturnicidae). *Molecular Phylogenetics and Evolution* 51:227–237.

Buckling, A., R. C. Maclean, M. A. Brockhurst, and N. Colegrave. 2009. The beagle in a bottle. *Nature* 457:824–829.

Buckling, A., and P. B. Rainey. 2002. The role of parasites in sympatric and allopatric host diversification. *Nature* 420:496–499.

Bull, C. M., and D. Burzacott. 2001. Temporal and spatial dynamics of a parapatric boundary between two Australian reptile ticks. *Molecular Ecology* 10:639–648.

Burgess, I. 2004. Human lice and their control. *Annual Review of Entomology* 49:457–481.

Burkett-Cadena, N. D. 2009. Morphological adapations of parasitic arthropods. In *Medical and Veterinary Entomology*, 2nd ed., edited by G. R. Mullen and L. A. Durden, 13–17. Academic Press/Elsevier Science, San Diego.

Burtt, E. H., Jr. 1986. An analysis of physical, physiological, and optical aspects of avian coloration with emphasis on wood-warblers. *Ornithological Monographs* 38:1–125.

Bush, A. O., J. C. Fernandez, G. W. Esch, and J. R. Seed. 2001. *Parasitism: The Diversity and Ecology of Animal Parasites*. Cambridge University Press, Cambridge.

Bush, S. E. 2004. *Evolutionary Ecology of Host Specificity in Columbiform Feather Lice*. PhD diss., University of Utah.

———. 2009. Does behavioural flexibility facilitate host switching by parasites? *Functional Ecology* 23:578–586.

Bush, S. E., and D. H. Clayton. 2006. The role of body size in host specificity: reciprocal transfer experiments with feather lice. *Evolution* 60:2158–2167.

Bush, S. E., C. W. Harbison, D. L. Slager, A. T. Peterson, R. D. Price, and D. H. Clayton. 2009a. Geographic variation in the community structure of lice on Western scrub-jays. *Journal of Parasitology* 95:10–13.

Bush, S. E., D. Kim, J. Lever, B. R. Moyer, D. H. Clayton. 2006a. Is melanin a defense against feather-feeding lice? *Auk* 123:153–161.

Bush, S. E., D. Kim, M. Reed, and D. H. Clayton. 2010. Evolution of cryptic coloration in ectoparasites. *American Naturalist* 176:529–535.

Bush, S. E., and J. R. Malenke. 2008. Host defence mediates interspecific competition in ectoparasites. *Journal of Animal Ecology* 77:558–564.

Bush, S. E., R. D. Price, and D. H. Clayton. 2009b. Descriptions of eight new species of feather lice in the genus *Columbicola* (Phthiraptera: Philopteridae), with a comprehensive world checklist. *Journal of Parasitology* 95:286–294.

Bush, S. E., M. Reed, and S. Maher. 2013. Influence of habitat fragmentation on ectoparasite diversity. *Biodiversity and Conservation* 22:1391–1404. doi:10.1007/s10531-013-0480-x.

Bush, S. E., A. N. Rock, S. L. Jones, J. R. Malenke, and D. H. Clayton. 2011. Efficacy of the LouseBuster, a new medical device for treating head lice (Anoplura: Pediculidae). *Journal of Medical Entomology* 48:67–72.

Bush, S. E., E. Sohn, and D. H. Clayton. 2006b. Ecomorphology of parasite attachment: experiments with feather lice. *Journal of Parasitology* 92:25–31.

Bush, S. E., S. M. Villa, T. J. Boves, D. Brewer, and J. R. Belthoff. 2012. Influence of bill and foot morphology on the ectoparasites of barn owls. *Journal of Parasitology* 98:256–261.

Busvine, J. R. 1946. On the pigmentation of the body louse *Pediculus humanus* L. *Proceedings of the Royal Entomological Society of London* 21:98–103.

———. 1976. *Insects, Hygiene and History*. Athlone Press.

———. 1978. Evidence from double infestations for the specific status of human head and body lice (Anoplura). *Systematic Entomology* 3:1–8.

Buxton, P. A. 1947. *The Louse*. Williams and Wilkins, Baltimore.

Caldwell, R. M., J. F. Schafer, L. E. Compton, and F. L. Patterson. 1958. Tolerance to cereal leaf rusts. *Science* 128:714–715.

Cameron, S. C., K. Yoshizawa, A. Mizukoshi, M. F. Whiting, K. P. Johnson. 2011. Mitochondrial genome deletions and minicircles are common in lice (Insecta: Phthiraptera). *BMC Genomics* 12:394.

Campbell, J. B. 1988. Arthropod induced stress in livestock. Veterinary Clinics of North America. *Food Animal Practice* 4:551–555.

Canestrari, D., D. Bolopo, T. C. Turlings, G. Röder, J. M. Marcos, and V. Baglione. 2014. From parasitism to mutualism: unexpected interactions between a cuckoo and its host. *Science* 343:1350–1352.

Cannon, S. M. 2010. Size correlations between sucking lice and their hosts including a test of Harrison's rule. Master's thesis, Georgia Southern University.

Charleston, M. A., and D. A. Robertson. 2002. Preferential host switching by primate lentiviruses can account for phylogenetic similarity with the primate phylogeny. *Systematic Biology* 51:528–535.

Chen, B. L., K. L. Haith, and B. A. Mullens. 2011. Beak condition drives abundance and grooming-mediated competitive asymmetry in a poultry ectoparasite community. *Parasitology* 138:748–757.

Chilton, N. B., F. Huby-Chilton, I. Beveridge, L. R. Smales, R. B. Gasser, and R. H. Andrews. 2011. Phylogenetic relationships of species within the tribe Labiostrongylinea (Nematoda: Cloacinidae) from Australian marsupials based on ribosomal DNA spacer sequence data. *Parasitology International* 60:381–387.

Christensen, N. O., P. Nansen, B. O. Fagbemi, and J. Monrad. 1987. Heterologous antagonistic and synergistic interactions between helminths and between helminths and protozoans in concurrent experimental infection of mammalian hosts. *Parasitology Research* 73:387–410.

Clark L., and J. R. Mason. 1985. Use of nest material as insecticidal and anti-pathogenic agents by the European starling. *Oecologia* 67:169–176.

———. 1988. Effect of biologically active plants used as nest material and the derived benefit to starling nestlings. *Oecologia* 77:174–180.

Clark, M. A., N. A. Moran, P. Baumann, J. J. Wernegreen. 2000. Cospeciation between bacterial endosymbionts (*Buchnera*) and a recent radiation of aphids (*Uroleucon*) and pitfalls of testing for phylogenetic congruence. *Evolution* 54:517–525.

Clay, T. 1949a. Piercing mouth-parts in the biting lice (Mallophaga). *Nature* 164:617.

———. 1949b. Some problems in the evolution of a group of ectoparasites. *Evolution* 3:279–299.

———. 1951. An introduction to the classification of the avian Ischnocera (Mallophaga): part I. *Transactions of the Royal Entomological Society of London* 102: 171–194.

———. 1969. A key to the genera of the Menoponidae:(Amblycera: Mallophaga: Insecta). *Bulletin British Museum (Natural History), Entomology* 24:3–26.

———. 1970. *The Amblycera (Phthiraptera: Insecta). Bulletin British Museum (Natural History), Entomology* 25:75–98.

———. 1974. The Phthiraptera (Insecta) parasitic on flamingoes (Phoenicopteridae: Aves). *Journal of Zoology* 172:483–490.

Clayton, A. L., K. F. Oakson, M. Gutin, et al. 2012. A novel human-infection derived bacterium provides insights into the evolutionary origins of mutualistic insect-bacterial symbioses. *PLoS Genetics* 8:e1002990.

Clayton, D. H. 1990a. Host specificity of *Strigiphilus* owl lice (Ischnocera: Philopteridae), with the description of new species and host associations. *Journal of Medical Entomology* 27:257–265.

———. 1990b. Mate choice in experimentally parasitized rock doves: lousy males lose. *American Zoologist* 30:251–262.

———. 1991a. Coevolution of avian grooming and ectoparasite avoidance. In *Bird-Parasite Interactions: Ecology, Evolution and Behaviour*, edited by J. E. Loye and M. Zuk, 258–289. Oxford University Press, Oxford.

———. 1991b. Influence of parasites on host sexual selection. *Parasitology Today* 7: 329–334.

Clayton, D. H., R. J. Adams, and S. E. Bush. 2008. Phthiraptera, the chewing lice. In *Parasitic Diseases of Wild Birds*, edited by C. T. Atkinson, N. J. Thomas, and D. B. Hunter, 515–526. Wiley-Blackwell, Ames.

Clayton, D. H., S. E. Al-Tamimi, and K. P. Johnson. 2003a. The ecological basis of coevolutionary history. In *Tangled Trees: Phylogeny, Cospeciation and Coevolution*, edited by R. D. M. Page, 310–341. University of Chicago Press, Chicago.

Clayton, D. H., S. E. Bush, B. M. Goates, and K. P. Johnson. 2003b. Host defense reinforces host-parasite coevolution. *Proceedings of the National Academy of Sciences USA* 100:15694–15699.

Clayton, D. H., S. E. Bush, and K. P. Johnson. 2004. The ecology of congruence: past meets present. *Systematic Biology* 53:165–173.

Clayton, D. H., and P. Cotgreave. 1994. Relationship of bill morphology to grooming behaviour in birds. *Animal Behavior* 47:195–201.

Clayton, D. H., R. D. Gregory, and R. D. Price. 1992. Comparative ecology of Neotropical bird lice (Insecta: Phthiraptera). *Journal of Animal Ecology* 61:781–795.

Clayton, D. H., and K. P. Johnson. 2001. What's bugging brood parasites? *Trends in Ecology and Evolution* 16:9–10.

———. 2003. Linking coevolutionary history to ecological process: doves and lice. *Evolution* 57:2335–2341.

Clayton, D. H., J. A. H. Koop, C. W. Harbison, B. R. Moyer, and S. E. Bush. 2010. How birds combat ectoparasites. *Open Ornithology Journal* 3:41–71.

Clayton, D. H., P. L. M. Lee, D. M. Tompkins, and E. D. Brodie III. 1999. Reciprocal natural selection on host-parasite phenotypes. *American Naturalist* 154:261–270.

Clayton, D. H., and J. Moore. 1997. Introduction. In *Host-Parasite Evolution: General Principles and Avian Models*, edited by D. H. Clayton and J. Moore, 1–6. Oxford University Press, Oxford.

Clayton, D. H., B. R. Moyer, S. E. Bush, et al. 2005. Adaptive significance of avian beak morphology for ectoparasite control. *Proceedings of the Royal Society of London B* 272:811–817.

Clayton, D. H., and R. D. Price. 1999. Taxonomy of New World *Columbicola* (Phthiraptera: Philopteridae) from the Columbiformes (Aves), with descriptions of five new species. *Annals of the Entomological Society of America* 92:675–685.

Clayton, D. H., R. D. Price, and R. D. M. Page. 1996. Revision of *Dennyus* (*Collodennyus*) lice (Phthiraptera: Menoponidae) from swiftlets, with descriptions of new taxa and a comparison of host-parasite relationships. *Systematic Entomology* 21:179–204.

Clayton, D. H., and D. M. Tompkins. 1994. Ectoparasite virulence is linked to mode of transmission. *Proceedings of the Royal Society of London B* 256:211–217.

Clayton, D. H., and D. M. Tompkins. 1995. Comparative effects of mites and lice on the reproductive success of rock doves (*Columba livia*). *Parasitology* 110:195–206.

Clayton, D. H., and J. G. Vernon. 1993. Common grackles anting with lime fruit and its effect on ectoparasites. *Auk* 110:951–952.

Clayton, D. H., and B. A. Walther. 2001. Influence of host ecology and morphology on the diversity of Neotropical bird lice. *Oikos* 94:455–467.

Clifford, C. M., J. F. Bell, G. J. Moore, and G. Raymond. 1967. Effects of limb disability on lousiness in mice. IV. Evidence of genetic factors in susceptibility to *Polyplax serrata*. *Experimental Parasitology* 20:56–67.

Clobert, J., E. Danchin, A. A. Dhondt, and J. D. Nichols, eds. 2001. *Dispersal*. Oxford University Press, Oxford.

Clobert, J., M. Baguette, T. G. Benton, and J. M. Bullock, eds. 2012. *Dispersal Ecology and Evolution*. Oxford University Press, Oxford.

Cobey, S., and M. Lipsitch. 2013. Pathogen diversity and hidden regimes of apparent competition. *American Naturalist* 181:12–24.

Cohen, S., M. T. Greenwood, and J. A. Fowler. 1991. The louse *Trinoton anserinum* (Amblycera: Phthiraptera), an intermediate host of *Sarconema eurycerca* (Filarioidea: Nematoda), a heartworm of swans. *Medical and Veterinary Entomology* 5:101–110.

Coley, P. D., and T. A. Kursar. 2014. On tropical forests and their pests. *Science* 343: 35–36.

Colles, A., L. H. Liow, and A. Prinzing. 2009. Are specialists at risk under environmental change? Neoecological, paleoecological and phylogenetic approaches. *Ecology Letters* 12:849–863.

Colwell, D. D., and C. Himsl-Rayner. 2002. *Linognathus vituli* (Anoplura: Linognathidae): population growth, dispersal and development of humoral immune responses in naive calves following induced infestations. *Veterinary Parasitology* 108:237–246.

Combes, C. 2001. *Parasitism: The Ecology and Evolution of Intimate Interactions*. University of Chicago Press, Chicago.

Condon, M. A., S. J. Scheffer, M. L. Lewis, R. Wharton, D. C. Adams, and A. A. Forbes. 2014. Lethal interactions between parasites and prey increase niche diversity in a tropical community. *Science* 343:1240–1244.

Connell, J. H. 1971. On the role of natural enemies in preventing competitive exclusion in some marine animals and in rain forest trees. In *Dynamics of*

Population, edited by P. J. Den Boer and G. R. Gradwell, 298–312. Centre for Agricultural Publishing and Documentation, Wageningen.

Cook, R. L., and L. S. Roberts. 1991. In vivo effects of putative crowding factors on development of *Hymenolepis diminuta*. *Journal of Parasitology* 77:21–25.

Corbet, G. B. 1956a. The life-history and host relations of a hippoboscid fly *Ornithomyia fringillina*. *Journal of Animal Ecology* 25:403–420.

———. 1956b. The phoresy of Mallophaga on a population of *Ornithomyia fringillina* Curtis (Dipt.: Hippoboscidae). *Entomologist's Monthy Magazine* 92:207–211.

Corbin, E., J. Vicente, M. P. Martin-Hernando, P. Acevedo, L. Perez-Rodriguez, and C. Gortazar. 2008. Spleen mass as a measure of immune strength in mammals. *Mammal Review* 38:108–115.

Cords, M. 1995. Predator vigilance costs of allogrooming in wild blue monkeys. *Behaviour* 132:559–569.

Corty, E. W., and J. M. Guardiani. 2008. Canadian and American sex therapists' perceptions of normal and abnormal ejaculatory latencies: how long should intercourse last? *Journal of Sexual Medicine* 5:1251–1256.

Cotgreave, P., and D. H. Clayton. 1994. Comparative analysis of time spent grooming by birds in relation to parasite load. *Behavior* 131:171–187.

Cott, H. B. 1940. *Adaptive coloration in animals*. Oxford University Press, Oxford.

Crespo, J. G., and N. J. Vickers. 2012. Antennal lobe organization in the slender pigeon louse, *Columbicola columbae* (Phthiraptera: Ischnocera). *Arthropod Structure & Development* 41:227–230.

Criscione, C. D., and M. S. Blouin. 2004. Life cycles shape parasite evolution: comparative population geneics of salmon trematodes. *Evolution* 58:198–202.

Criscione, C. D., R. Poulin, and M. S. Blouin. 2005. Molecular ecology of parasites: elucidating ecological and microevolutionary processes. *Molecular Ecology* 14: 2247–2257.

Crispo, E. 2007. The Baldwin effect and genetic assimilation: revisiting two mechanisms of evolutionary change mediated by phenotypic plasticity. *Evolution* 61:2469–2479.

Croll, D. A., and E. McLaren. 1993. Diving metabolism and thermoregulation in common and thick-billed murres. *Journal of Comparative Physiology B: Biochemical, Systemic, and Environmental Physiology* 163:160–166.

Cruz, S., and M. P. Martín Mateo. 2009. Scanning electron microscopy of legs of two species of sucking lice (Anoplura: Phthiraptera). *Micron* 40:401–408.

Currie, C. R., B. Wong, A. E. Stuart, et al. 2003. Ancient tripartite coevolution in the attine ant-microbe symbiosis. *Science* 299:386–388.

Curtis, V., and A. Biran. 2001. Dirt, disgust, and disease: is hygiene in our genes? *Perspectives in Biology and Medicine* 44:17–31.

Cushman, H., and T. G. Whitham. 1989. Conditional mutualism in a membracid-ant association: temporal, age-specific, and density-dependent effects. *Ecology* 70: 1040–1047.

Cuthill, J. H., and M. A. Charleston. 2013. A simple model explains the dynamics of preferential host switching among mammal RNA viruses. *Evolution* 67:980–990.

Czirják, G. Á., P. L. Pap, C. I. Vágási, et al. 2013. Preen gland removal increases plumage bacterial load but not that of feather-degrading bacteria. *Naturwissenschaften* 100:145–151.

Da Silva, A. S., L. S. Lopes, J. D. S. Diaz, A. A. Tonin, L. M. Stefani, and D. N. Araújo. 2013a. Lice outbreak in buffaloes: evidence of *Anaplasma marginale* transmission by sucking lice *Haematopinus tuberculatus. Journal of Parasitology* 99:546–547.

Da Silva, A. S., A. A. Tonin, and L. S. Lopes. 2013b. Outbreak of lice in horses: epidemiology, diagnosis, and treatment. *Journal of Equine Veterinary Science* 33: 530–532.

Dabert, J. 2003. The feather mite family Syringobiidae Trouessart, 1896 (Acari, Astigmata, Pterolichoidea). II. Phylogeny and host-parasite evolutionary relationships. *Acta Parasitologica* 48:S185–233.

Dabert, J., M. Daber, and S. V. Mironov. 2001. Phylogeny of feather mite subfamily Avenzoariinae (Acari: Analgoidea: Avenzariidae) inferred from combined analyses of molecular and morphological data. *Molecular Phylogenetics and Evolution* 20: 124–135.

Dale, C., and N. A. Moran. 2006. Molecular interactions between bacterial symbionts and their hosts. *Cell* 126:453–465.

Darolova, A., H. Hoi, J. Kristofik, and C. Hoi. 2001. Horizontal and vertical ectoparasite transmission of three species of Mallophaga, and individual variation in European bee-eaters (*Merops apiaster*). *Journal of Parasitology* 87:256–262.

Darwin, C. 1837. Notebook B [Transmutation of species (1837–1838)]. Darwin Online. http://darwin-online.org.uk.

Darwin, C. 1859. *On the origin of species by means of natural selection.* John Murray, London.

———. 1862. *On the various contrivances by which British and foreign orchids are fertilized by insects, and on the good effects of intercrossing.* John Murray, London.

———. 1871. *Charles Darwin's Zoology Notes and Specimen Lists from H. M. S. Beagle.* Cambridge University Press, Cambridge. Published online 2005.

———. 1888. *The descent of man and selection in relation to sex.* 2nd ed. John Murray, London.

David, P., and P. Heeb. 2009. Parasites and sexual selection. In *Ecology and Evolution of Parasitism*, edited by F. Thomas, J. Guegan, and F. Ranaud, 31–47. Oxford University Press, Oxford.

Davies, N. B. 2000. *Cuckoos, Cowbirds, and Other Cheats.* T & A. D. Poyser, London.

Davis, D. P., and R. E. Williams. 1986. Influence of hog lice, *Haematopinus suis*, on blood components, behavior, weight gain and feed efficiency of pigs. *Veterinary Parasitology* 22:307–314.

Dawkins, R. 2005. The ancestor's tale: a pilgrimage to the dawn of evolution. Houghton Mifflin Harcourt, New York.

Dawkins, R., and J. R. Krebs. 1979. Arms races between and within species. *Proceedings of the Royal Society of London B* 205:489–511.

de Meeûs, T., and F. Renaud. 2002. Parasites within the new phylogeny of eukaryotes. *Trends in Parasitology* 18:247–251.

de Vienne, D. M., G. Refrégier, M. López-Villavicencio, A. Tellier, M. E. Hood, and T. Giraud. 2013. Cospeciation vs host-shift speciation: methods for testing, evidence from natural associations and relation to coevolution. *New Phytologist* 198:347–385.

de Vienne, D. M., T. Giraud, and J. A. Shykoff. 2007. When can host shifts produce congruent host and parasite phylogenies? A simulation approach. *Journal of Evolutionary Biology* 20:1428–1438.

del Hoyo, J., A. Elliot, and J. Sargatal, eds. 1997. *Handbook of the Birds of the World.* Vol 4. *Sandgrouse to Cuckoos.* Lynx Edicions, Barcelona.

Demastes, J. W., M. S. Hafner, D. J. Hafner, and T. A. Spradling. 1998. Pocket gophers and chewing lice: a test of the maternal transmission hypothesis. *Molecular Ecology* 7:1065–1069.

Demastes, J. W., T. A. Spradling, and M. S. Hafner. 2003. The effects of spatial and temporal scale on analyses of cophylogeny. In *Tangled Trees: Phylogeny, Cospeciation, and Coevolution,* edited by R. D. M. Page, 221–239. University of Chicago Press, Chicago.

Demastes, J. W., T. A. Spradling, M. S. Hafner, G. R. Spies, D. J. Hafner, and J. E. Light. 2012. Cophylogeny on a fine scale: *Geomydoecus* chewing lice and their pocket gopher hosts, *Pappogeomys bulleri. Journal of Parsitology* 98:262–270.

Dentzien-Dias, P. C., G. Poinar Jr., A. E. Q. de Figueiredo, et al. 2013. Tapeworm eggs in a 270 million-year-old shark coprolite. *PLoS ONE* 8:e55007. doi:10.1371/journal. pone.0055007.

DeVries, P., and G. O. Poinar. 1997. Ancient butterfly-ant symbiosis: direct evidence from Dominican amber. *Proceedings of the Royal Society of London B* 264:1137–1140.

Dietl, G. P. 2003. Coevolution of a marine gastropod predator and its dangerou bivalve prey. *Biological Journal of the Linnean Society* 80:409–436.

Dietl, G. P., S. R. Durham, and P. H. Kelley. 2010. Shell repair as a reliable indicator of bivalve predation by shell-wedging gastropods in the fossil record. *Palaeogeography, Palaeoclimatology, Palaeoecology* 296:174–184.

Dietl, G. P., and P. H. Kelly. 2002. The fossil record of predator-prey arms races: coevolution and escalation hypotheses. In *The fossil record of predation,* edited by M. Kowalewski and P. H. Kelley. *Paleontological Society Papers* 8:353–374.

Dik, B. 2006. Erosive stomatitis in a white pelican (*Pelecanus onocrotalus*) caused by *Piagetiella titan* (Mallophaga: Menoponidae). *Journal of Veterinary Medicine B* 53: 153–154.

Dillon, R. J., C. T. Vennard, A. Buckling, and A. K. Charnley. 2005. Diversity of locust gut bacteria protects against pathogen invasion. *Ecology Letters* 8:1291–1298.

Dobson, A. P. 1985. The population dynamics of competition between parasites. *Parasitology* 91:317.

Dobzhansky, T. 1948. Darwin's finches and evolution. *Ecology* 29:219–220.

Doucet, S. M., and R. Montgomerie. 2003a. Multiple sexual ornaments in satin bowerbirds: ultraviolet plumage and bowers signal different aspects of male quality. *Behavioral Ecology* 14:503–509.

———. 2003b. Structural plumage colour and parasites in satin bowerbirds

Ptilonorhynchus violaceus: implications for sexual selection. *Journal of Avian Biology* 34:237–242.

Douglas, H. D., III. 2013. Colonial seabird's paralytic perfume slows lice down: an opportunity for parasite-mediated selection? *International Journal for Parasitology* 43:399–407.

Douglas, H. D., III, J. R. Malenke, and D. H. Clayton. 2005. Is the citrus-like plumage odorant of crested auklets (*Aethia cristatella*) a defense against lice? *Journal of Ornithology* 146:111–115.

Downton, M., and A. D. Austin. 1995. Increased genetic diversity in mitochondrial genes is correlated with the evolution of parasitism in the Hymenoptera. *Journal of Molecular Evolution* 41:958–965.

Duffy, M. A., and S. E. Forde. 2009. Ecological feedbacks and the evolution of resistance. *Journal of Animal Ecology* 78:1106–1112.

Dumbacher, J. P. 1999. Evolution of toxicity in pitohuis. I. Effects of homobatrachotoxin on chewing lice (order Phthiraptera). *Auk* 116:957–963.

Dumbacher, J. P., B. M. Beehler, T. F. Spande, H. M. Garraffo, and J. W. Daly. 1992. Homobatrachotoxin in the genus *Pitohui*: chemical defense in birds? *Science* 258: 799–801.

Dumbacher, J. P., and S. Pruett-Jones. 1996. Avian chemical defense. *Current Ornithology* 13:137–174.

Dumbacher J. P., T. F. Spande, and J. W. Daly. 2000. Batrachotoxin alkaloids from passerine birds: a second toxic bird genus (*Ifrita kowaldi*) from New Guinea. *Proceedings of the National Academy of Sciences USA* 97:12970–12975.

Dunbar, R. I. 1993. Coevolution of neocortical size, group size and language in humans. *Behavioral and Brain Sciences* 16:681–693.

———. 1996. *Grooming, Gossip and the Evolution of Language.* Faber and Faber, London.

———. 2003. The social brain: mind, language, and society in evolutionary perspective. *Annual Review of Anthropology* 32:163–181.

Dunn, R., N. Harris, R. Colwell, L. Koh, and N. Sodhi. 2009. The sixth mass coextinction: are most endangered species parasites and mutualists? *Proceedings of the Royal Society of London B* 276:3037–3045.

Dunning, J. B. 1993. *CRC Handbook of Avian Body Masses.* CRC, Boca Raton.

Dunson, W. A., and J. Travis. 1994. Patterns in the evolution of physiological specialization in salt-marsh animals. *Estuaries and Coasts* 17:102–110.

Durden, L. A. 1983. Sucking louse (*Hoplopleura erratica*: Insecta, Anoplura) exchange between individuals of a wild population of eastern chipmunks, *Tamias striatus*, in central Tennessee, USA. *Journal of Zoology of London* 201:117–123.

———. 1990. Phoretic relationships between sucking lice (Anoplura) and flies (Diptera) associated with humans and livestock. *Entomologist* 109:191–192.

———. 2001. Lice (Phthiraptera). In *Parasitic Diseases of Wild Mammals*, edited by W. M. Samuel, M. J. Pybus, and A. A. Kocan, 3–17. Iowa State University Press, Ames.

———. 2005. Lice (Phthiraptera). In *The Biology of Disease Vectors*, 2nd ed., edited by

W. C. Marquardt, W. C. Black IV, J. Freier, et al., 67–75. Elsevier/Academic Press, San Diego.

Durden, L. A., and J. E. Lloyd. 2009. Lice (Phthiraptera). In *Medical and Veterinary Entomology*, 2nd ed., edited by G. R. Mullen and L. A. Durden, 59–82. Academic Press/Elsevier Science, San Diego.

Durden, L. A., and G. G. Musser. 1994. The sucking lice (Insecta: Anoplura) of the world: a taxonomic checklist with records of mammalian hosts and geographic distributions. *Bulletin of the American Museum of Natural History* 218:1–90.

Dusbábek, F., and V. Skarkova-Spakova. 1988. Acquired resistance of pigeons to *Argas polonicus* larvae. *Folia Parasitologica* 35:77–84.

Dyke, G. J., and M. van Tuinen. 2004. The evolutionary radiation of modern birds (Neornithes): reconciling molecules, morphology and the fossil record. *Zoological Journal of the Linnean Society* 141:153–177.

Ebert, D. 1998. Experimental evolution of parasites. *Science* 282:1432–1435.

Eens, M., E. Van Duyse, and L. Berghman. 2000. Shield characteristics are testosterone-dependent in both male and female moorhens. *Hormones and Behavior* 37:126–134.

Ehrlich, P. R. and P. H. Raven. 1964. Butterflies and plants: a study in coevolution. *Evolution* 18:586–608.

Eichler, W. 1936. Die biologie der Federlinge. *Journal of Ornithology* 84:471–505.

———. 1942. Die Entfaltungsregel und andere Gesetzmäßigkeiten in den parasitogenetischen Beziehungen der Mallophagen und anderer standiger Parasiten zu ihren Wirten. *Zoologischer Anzeiger* 137:77–83.

———. 1963. Arthropoda. Insecta. Phthiraptera 1. Mallophaga. *Klassen und Ordnungen des tierreichs*. H. G. Bronns, ed. 111. Insecta. 7b Phthiraptera. Leipzig.

Ericson, P. G. P., C. L. Anderson, T. Britton, et al. 2006. Diversification of Neoaves: integration of molecular sequence data and fossils. *Biological Letters* 2:543–547.

Esch, G. W., A. O. Bush, and J. M. Aho, eds. 1990. *Parasite Communities: Patterns and Processes*. Chapman and Hall, London.

Eveleigh, E. S., and W. Threlfall. 1976. Population dynamics of lice (Mallophaga) on auks (Alcidae) from Newfoundland. *Canadian Journal of Zoology* 54:1694–1711.

Ewald, P. W. 1994. *Evolution of Infectious Disease*. Oxford University Press, New York.

Fahrenholz, V. H. 1915. Läuse verschiedener Menschenrassen. *Zeitschrift für Morphologie und Anthropologie* 17:591–602.

Fairn, E. R., N. R. McLellan, and D. Shutler. 2012. Are lice associated with ring-billed gull chick immune responses? *Waterbirds* 35:164–169.

Falconer, D. S. 1981. *Introduction to Quantitative Genetics*. 2nd ed. Longman, New York.

Falótico, T., M. B. Labruna, M. P. Verderane, B. D. De Resende, P. Izar, and E. B. Ottoni. 2007. Repellent efficacy of formic acid and the abdominal secretion of carpenter ants (Hymenoptera: Formicidae) against Amblyomma ticks (Acari: Ixodidae). *Journal of Medical Entomology* 44:718–721.

Farrell, B. D. 1998. "Inordinate fondness" explained: why are there so many beetles? *Science* 281:555–559.

Farrell, B. D., and C. Mitter. 1994. Adaptive radiation in insects and plants: time and opportunity. *American Zoologist* 34:57–69.

Farrell, B. D., and A. S. Sequeira. 2004. Evolutionary rates in the adaptive radiation of beetles on plants. *Evolution* 58:1984–2001.

Favret, C., and D. J. Voegtlin. 2004. Speciation by host-switching in pinyon *Cinera* (Insecta: Hemiptera: Aphididae). *Molecular Phylogenetics and Evolution* 32:139–151.

Feingold, B. F., and E. Benjamini. 1961. Allergy to flea bites: clinical and experimental observations. *Annals of Allergy* 19:1275.

Felso, B., and L. Rózsa. 2006. Reduced taxonomic richness of lice (Insecta: Phthiraptera) in diving birds. *Journal of Parasitology* 92:867–869.

———. 2007. Diving behavior reduces genera richness of lice (Insecta: Phthiraptera) of mammals. *Acta Parasitologica* 52:82–85.

Fenton, A., and S. E. Perkins. 2010. Applying predator-prey theory to modelling immune-mediated, within-host interspecific parasite interactions. *Parasitology* 137:1027.

Ferrari, N., I. M. Cattadori, A. Rizzoli, and P. J. Hudson. 2009. *Heligmosomoides polygyrus* reduces infestation of *Ixodes ricinus* in free-living yellow-necked mice, *Apodemus flavicollis*. *Parasitology* 136:305–316.

Fey, A. J., D. L. Oliver, and M. B. Williams. 1997. Theft of nesting material involving honeyeaters (Meliphagidae). *Corella* 21:119–123.

Filchak, K. E., J. B. Roethele, and J. L. Feder. 2000. Natural selection and sympatric divergence in the apple maggot *Rhagoletis pomonella*. *Nature* 407:739–742.

Folstad, I., and A. J. Karter. 1992. Parasites, bright males, and the immunocompetence handicap. *American Naturalist* 139:603–622.

Forde, S. E., J. N. Thompson, and B. J. Bohannan. 2004. Adaptation varies through space and time in a coevolving host–parasitoid interaction. *Nature* 431:841–844.

Forister, M. L., L. A. Dyer, M. S. Singer, J. O. Stireman III, and J. T. Lill. 2012. Revisiting the evolution of ecological specialization, with emphasis on insect-plant interactions. *Ecology* 93:981–991.

Foster, M. S. 1969. Synchronised life cycles in the orange-crowned warbler and its mallophagan parasites. *Ecology* 50:315–323.

Fowler, J. A., and L. R. Williams. 1985. Population dynamics of Mallophaga and Acari on reed buntings occupying a communal winter roost. *Ecological Entomology* 10: 377–383.

Fox, L. R. 1975. Cannibalism in natural populations. *Annual Review of Ecology and Systematics* 6:87–106.

Freed, L. A., R. C. Cann, and G. R. Bodner. 2008a. Incipient extinction of a major population of the Hawaii akepa owing to introduced species. *Evolutionary Ecology Research* 10:931–965.

Freed, L. A., M. C. Medeiros, and G. R. Bodner. 2008b. Explosive icrease in ectoparasites in Hawaiian forest birds. *Journal of Parasitology* 94:1009–1021.

Friggens, M. M., and J. H. Brown. 2005. Niche partitioning in the cestode communities of two elasmobranchs. *Oikos* 108:76–84.

Fritz, R. S., C. Moulia, and G. Newcombe. 1999. Resistance of hybrid plants and animals to herbivores, pathogens, and parasites. *Annual Review of Ecology and Systematics* 30:565–591.

Fritz, R. S., and E. L. Simms, eds. 1992. *Plant Resistance to Herbivores and Pathogens*. University of Chicago Press, Chicago

Fukatsu, T., R. Koga, W. A. Smith, et al. 2007. Bacterial endosymbiont of the slender pigeon louse *Columbicola columbae*, allied to endosymbionts of grain weevils and tsetse flies. *Applied and Environmental Microbiology* 73:6660–6669.

Futuyma, D. J. 2010. How species affect each other's evolution. *Evolution: Education and Outreach* 3:3–5.

———. 2013. *Evolution*. 3rd ed. Sinauer Associates, Sunderland.

Futuyma, D. J., and A. A. Agrawal. 2009. Macroevolution and the biological diversity of plants and herbivores. *Proceedings of the National Academy of Sciences USA* 106: 18054–18061.

Futuyma, D., and G. Moreno. 1988. The evolution of ecological specialization. *Annual Review of Ecology and Systematics* 19:207–233.

Gaba, S., and D. Ebert. 2009. Time-shift experiments as a tool to study antagonistic coevolution. *Trends in Ecology & Evolution* 24:226–232.

Gandon, S. 2002. Local adaptation and the geometry of host-parasite coevolution. *Ecology Letters* 5:246–256.

Gandon, S., A. Buckling, E. Decaestecker, and T. Day. 2008. Host-parasite coevolution and patterns of adaptation across time and space. *Journal of Evolutionary Biology* 21:1861–1866.

Gandon, S., and Y. Michalakis. 2002. Local adaptation, evolutionary potential and host-parasite coevolution: interactions between migration, mutation, population size and generation time. *Journal of Evolutionary Biology* 15:451–462.

Garamszegi, L. Z. 2005. Bird song and parasites. *Behavioral Ecology and Sociobiology* 59:167–180.

Garamszegi, L. Z., D. Heylen, A. P. Møller, M. Eens, and F. De Lope. 2005. Age-dependent health status and song characteristics in the barn swallow. *Behavioral Ecology* 16:580–591.

Garrido, E., G. Andraca-Gómez, and J. Fornoni. 2012. Local adaptation: simultaneously considering herbivores and their host plants. *New Phytologist* 193:445–453.

Gavrilets, S., and J. B. Losos. 2009. Adaptive radiation: contrasting theory with data. *Science* 323:732–737.

Gerardo, N. M., U. G. Mueller, and C. R. Currie. 2006. Complex host-pathogen coevolution in the *Apterostigma* fungus-growing ant-microbe symbiosis. *BMC Evolutionary Biology* 6:88.

Gibbs, D., E. Barnes, and J. Cox. 2001. *Pigeons and Doves: A Guide to Pigeons and Doves of the World*. Yale University Press, New Haven.

Gillespie, J. H. 2001. Is the population size of a species relevant to its evolution? *Evolution* 55:145–156.

Gillespie, J. M., and M. J. Frenkel. 1974. The diversity of keratins. *Comparative Biochemical Physiology* 47B:339–346.

Gillespie, R. G. 2004. Community assembly through adaptive radiation in Hawaiian spiders. *Science* 303:356–359.

Gillespie, R. G. 2009. Adaptive radiation. In *The Princeton Guide to Ecology*, edited by S. A. Levin, S. R. Carpenter, H. Charles, et al., 143–152. Princeton University Press, Princeton.

Giorgi, M. S., R. Arlettaz, P. Christe, and P. Vogel. 2001. The energetic grooming costs imposed by a parasitic mite (*Spinturnix myoti*) upon its bat host (*Myotis myotis*). *Proceedings of the Royal Society of London B* 268:2071–2075.

Glor, R. E. 2010. Phylogenetic insights on adaptive radiation. *Annual Review of Ecology, Evolution, and Systematics* 41:251–270.

Goater, T. M., C. P. Goater, and G. W. Esch. 2013. *Parasitism: The Diversity and Ecology of Animal Parasites*. 2nd ed. Cambridge University Press, New York.

Goldstein, D. L. 1988. Estimates of daily energy expenditure in birds: the time-energy budget as an integrator of laboratory and field studies. *American Zoologist* 28: 829–844.

Gómez, P., and A. Buckling. 2011. Bacteria-phage antagonistic coevolution in soil. *Science* 332:106–109.

Gomulkiewicz, R., D. M. Drown, M. F. Dybdahl, et al. 2007. Do's and don'ts of testing the geographic mosaic theory of coevolution. *Heredity* 98:249–258.

Gorb, S. 2001. *Attachment Devices of Insect Cuticle*. Kluwer Academic, London.

Gould, S. J. 1993. *Eight Little Piggies; Reflections in Natural History*. W. W. Norton, New York.

Grant, P. R., and B. R. Grant. 2008. *How and Why Species Multiply: The Radiation of Darwin's Finches*. Princeton University Press, Princeton.

Greene, H. W. 2005. Organisms in nature as a central focus for biology. *Trends in Ecology and Evolution* 20:23–27.

Greenslade, P. J. M. 1983. Adversity selection and the habitat templet. *American Naturalist* 122:352–365.

Greer, J. M., and M. R. Capecchi. 2002. *Hoxb8* is required for normal grooming behavior in mice. *Neuron* 33:23–34.

Greischar, M. A., and B. Koskella. 2007. A synthesis of experimental work on parasite local adaptation. *Ecology Letters* 10:418–434.

Grimaldi, D. A., and M. S. Engel. 2006. Fossil Liposcelididae and the lice ages (Insecta: Psocodea). *Proceedings of the Royal Society of London B* 273:625–633. doi: 10.1098/rspb.2005.3337.

Grossi, A. A., B. J. Sharanowski, and T. D. Galloway. 2014. *Anatoecus* species (Phthiraptera: Philopteridae) from Anseriformes in North America and taxonomic status of *Anatoecus dentatus* and *Anatoecus iceterodes*. *Canadian Entomologist* 146:598–608.

Grueter, C. C., A. Bissonnette, K. Isler, and C. P. van Schaik. 2012. Grooming and group cohesion in primates: implications for the evolution of language. *Evolution and Human Behavior* 34:61–68.

Guilhem, R., A. Simková, S. Morand, and S. Gourbiere. 2012. Within-host competition and diversification of macro-parasites. *Journal of the Royal Society Interface* 9:2936–2946.

Gupta, N., V. Khan, S. Kumar, S. Saxena, A. Rashmi, and A. K. Saxena. 2009. Eggshell morphology of selected Indian bird lice (Phthiraptera: Amblycera and Ischnocera). *Entomological News* 120:327–336.

Hackett, S. J., R. T. Kimball, S. Reddy, et al. 2008. A phylogenomic study of birds reveals their evolutionary history. *Science* 320:1763–1768.

Hafner, M. S., J. W. Demastes, D. J. Hafner, T. A. Spradling, P. D. Sudman, and S. A. Nadler. 1998. Age and movement of a hybrid zone: implications for dispersal distance in pocket gophers and their chewing lice. *Evolution* 52:278–282.

Hafner, M. S., J. W. Demastes, T. A. Spradling, and D. L. Reed. 2003. Cophylogeny between pocket gophers and chewing lice. In *Tangled Trees: Phylogeny, Cospeciation and Coevolution,* edited by R. D. M. Page, 195–220. University of Chicago Press, Chicago.

Hafner, M. S., and S. A. Nadler. 1988. Phylogenetic trees support the coevolution of parasites and their hosts. *Nature* 332:258–259.

Hafner, M. S., P. D. Sudman, F. X. Villablanca, T. A. Spradling, J. W. Demastes, and S. A. Nadler. 1994. Disparate rates of molecular evolution in cospeciating hosts and parasites. *Science* 265:1087–1090.

Hagelin, J. C., and I. L. Jones. 2007. Bird odors and other chemical substances: a defense mechanism or overlooked mode of intraspecific communication? *Auk* 124:741–761.

Hahn, D. C., R. D. Price, and P. C. Osenton. 2000. Use of lice to identify cowbird hosts. *Auk* 117:947–955.

Haldane, J. B. S. 1949. Disease and evolution. Supplement to *La Ricerca Scientifica* 19: 68–76.

Hamilton, W. D., and M. Zuk. 1982. Heritable true fitness and bright birds: a role for parasites? *Science* 218:384–387.

Hanifin, C. T., E. D. Brodie Jr., and E. D. Brodie III. 2008. Phenotypic mismatches reveal escape from arms-race coevolution. *PLoS Biology* 6:e60.

Hanski, I. 1998. Metapopulation dynamics. *Nature* 396:41–49.

Harbison, C. W., and R. M. Boughton. 2014. Thermo-orientation and the movement of feather-feeding lice on hosts. *Journal of Parasitology* 100:433–441.

Harbison, C. W., and D. H. Clayton. 2011. Community interactions govern host switching with implications for host-parasite coevolutionary history. *Proceedings of the National Acadamy of Sciences USA* 108:9525–9529.

Harbison, C. W., S. E. Bush, J. R. Malenke, and D. H. Clayton. 2008. Comparative transmission dynamics of competing parasite species. *Ecology* 89:3186–3194.

Harbison, C. W., M. V. Jacobsen, and D. H. Clayton. 2009. A hitchhiker's guide to parasite transmission: the phoretic behaviour of feather lice. *International Journal for Parasitology* 39:569–575.

Harrison, L. 1915. Mallophaga from Apteryx, and their significance; with a note on the genus Rallicola. *Parasitology* 8:88–100.

Hart, B. L. 1997. Behavioural defence. In *Host-Parasite Evolution: General Principles and Avian Models*, edited by D. H. Clayton and J. Moore, 59–77. Oxford University Press, Oxford.

Hart, B. L., L. A. Hart, M. S. Mooring, and R. Olubayo. 1992. Biological basis of grooming behaviour in antelope: the body-size, vigilance and habitat principles. *Animal Behavior* 44:615–631.

Harvey, P., and A. Keymer. 1991. Comparing life histories using phylogenies. *Philosophical Transactions of the Royal Society B* 332:31–39.

Hassanin, A. 2006. Phylogeny of Arthropoda inferred from mitochondrial sequences: strategies for limiting the misleading effects of multiple changes in pattern and rates of substitution. *Molecular Phylogenetics and Evolution* 38:100–166.

Hatcher, M. J., and A. M. Dunn. 2011. *Parasites in Ecological Communities: From Interactions to Ecosystems*. Cambridge University Press, New York.

He, T., J. G. Pausas, C. M. Belcher, D. W. Schwilk, and B. B. Lamont. 2012. Fire-adapted traits of *Pinus* arose in the fiery Cretaceous. *New Phytologist* 194:751–759.

Healy, W. M., and J. W. Thomas. 1973. Effects of dusting on plumage of Japanese quail. *Wilson Bulletin* 85:442–448.

Heeb, P. I., I. Werner, M. Kölliker, and H. Richner. 1998. Benefits of induced host responses against an ectoparasite. *Proceedings of the Royal Society of London B* 265: 51–56.

Hembry, D. H., J. B. Yoder, and K. R. Goodman. 2014. Coevolution and the diversification of life. *American Naturalist* 184:425–438.

Hemmes, R. B., A. Alvarado, and B. L. Hart. 2002. Use of California bay foliage by wood rats for possible fumigation of nest-borne ectoparasites. *Behavioral Ecology* 13:381–385.

Herbert, P. D. N., and C. J. Emery. 1990. The adaptive significance of cuticular pigmentation in *Daphnia*. *Functional Ecology* 4:703–710.

Herron, J. C., and S. Freeman. 2014. *Evolutionary Analysis*. 5th ed. Pearson Education, Upper Saddle River, NJ.

Hillegass, M. A., J. M. Waterman, and J. D. Roth. 2008. The influence of sex and sociality on parasite loads in an African ground squirrel. *Behavioral Ecology* 19: 1006–1011.

Hillgarth, N. 1996. Ectoparasite transfer during mating in ring-necked pheasants *Phasianus colchicus*. *Journal of Avian Biology* 27:260–262.

Hink, W. F., and B. J. Fee. 1986. Toxicity of D-limonene, the major component of citrus peel oil, to all life stages of the cat flea, *Ctenocephalides felis* (Siphonaptera: Pulicidae). *Journal of Medical Entomology* 23:400–404.

Hirose, S., S. Tanda, L. Kiss, B. Grigaliunaite, M. Havrylenko, and S. Takamatsu. 2005. Molecular phylogeny and evolution of the maple powdery mildew (*Sawadaea, Erysiphaceae*) inferred from nuclear rDNA sequences. *Mycological Research* 109:912–922.

Hirschfelder, A. D., and W. Moore. 1919. Clinical studies on the effects of louse bites: *Pediculus corporis*. *Archives of Internal Medicine* 23:419–430.

Hobæk, A., and H. G. Wolf. 1991. Ecological genetics of Norwegian *Daphnia*. II. Distribution of *Daphnia longispina* genotypes in relation to short-wave radiation and water colour. *Hydrobiologia* 225:229–243.

Hoberg, E. P., D. R. Brooks, and D. Siegel-Causey. 1997. Host parasite co-speciation: history, principles and prospects. In *Host-Parasite Evolution: General Principles and Avian Models*, edited by D. H. Clayton and J. Moore, 212–235. Oxford University Press, Oxford.

Hodgdon, H. E., K. S. Yoon, D. J. Previte, et al. 2010. Determination of knockdown resistance allele frequencies in global human head louse populations using the serial invasive signal amplification reaction. *Pest Management Science* 66:1031–1040.

Hoeck, P. E., and L. F. Keller. 2012. Inbreeding, immune defence and ectoparasite load in different mockingbird populations and species in the Galápagos Islands. *Journal of Avian Biology* 43:423–434.

Hoeksema, J. D. 2012. Geographic mosaics of coevolution. *Nature Education Knowledge* 3:19.

Hoeksema, J. D., and S. E. Forde. 2008. A meta-analysis of factors affecting local adaptation between interacting species. *American Naturalist* 171:275–290.

Hoekstra, H. E. 2010. From mice to molecules: the genetic basis of color adaptation. In *In the Light of Evolution: Essays from the Laboratory and Field*, edited by J. B. Losos, 277–295. Roberts, Greenwood Village, CO.

Hoi, H., J. Krištofík, A. Darolová, and C. Hoi. 2012. Experimental evidence for costs due to chewing lice in the European bee-eater (*Merops apiaster*). *Parasitology* 139: 53–59.

Holdenried, R., F. C. Evans, and D. S. Longanecker. 1951. Host-parasite-disease relationships in a mammalian community in the central coast range of California. *Ecological Monographs* 21:1–18.

Holmes, J. C. 1973. Site selection by parasitic helminths: interspecific interactions, site segregation, and their importance to the development of helminth communities. *Canadian Journal of Zoology* 51:333–347.

Holt, R. D. 1977. Predation, apparent competition, and the structure of prey communities. *Theoretical Population Biology* 12:197–129.

Hopkins, G. H. E. 1942. The Mallophaga as an aid to the classification of birds. *Ibis* 84:94–106.

———. 1949. The host-associations of the lice of mammals. *Proceedings of the Zoological Society of London* 119:387–604.

Hornok, S., R. Hofmann-Lehmann, I. G. Fernández de Mera, et al. 2010. Survey on blood-sucking lice (Phthiraptera: Anoplura) of ruminants and pigs with molecular detection of *Anaplasma* and *Rickettsia* spp. *Veterinary Parasitology* 174: 355–358.

Hoyle, W. L. 1938. Transmission of poultry parasites by birds with special reference to the "English" or house sparrow and chickens. *Transactions of the Kansas Academy of Science* 41:379–384.

Hu, S., D. L. Dilcher, D. M. Jarzen, and D. W. Taylor. 2008. Early steps of angiosperm-pollinator coevolution. *Proceedings of the National Academy of Sciences USA* 105: 240–245.

Huber, S. K., J. P. Owen, J. A. H. Koop, et al. 2010. Ecoimmunity in Darwin's finches: invasive parasites trigger acquired immunity in the medium ground finch (*Geospiza fortis*). *PLoS ONE* 5:e8605.

Hudault, S., J. Guignot, and A. L. Servin. 2001. *Escherichia coli* strains colonising the gastrointestinal tract protect germfree mice against *Salmonella typhimurium* infection. *Gut* 49:47–55.

Hughes, J., M. Kennedy, K. P. Johnson, R. L. Palma, and R. D. M. Page. 2007. Multiple cophylogenetic analyses reveal frequent cospeciation between pelecaniform birds and *Pectinopygus* lice. *Systematic Biology* 56:232–251.

Hughes, J., and R. D. M. Page. 2007. Comparative tests of ectoparasite species richness in seabirds. *BMC Evolutionary Biology* 7:227.

Hugot, J. P. 1999. Primates and their pinworm parasites: the Cameron hypothesis revisited. *Systematic Biology* 48:523–546.

———. 2003. New evidence for hystricognath rodent monophyly from the phylogeny of their pinworms. In *Tangled Trees: Phylogeny, Cospeciation and Coevolution,* edited by R. D. M. Page, 144–173. University of Chicago Press, Chicago.

Hugot, J. P., J. P. Gonzalez, and C. Denys. 2001. Evolution of the old work Arenaviridae and their rodent hosts: generalized host-transfer or association by descent? *Infection, Genetics and Evolution* 1:13–20.

Humphrey-Smith, I. 1989. The evolution of phylogenetic specificity among parasitic organisms. *Parasitology Today* 5:385–387.

Humphries, D. A. 1967. Function of combs in ectoparasites. *Nature* 215:319.

Huron, D. 2001. Is music an evolutionary adaptation? *Annals of the New York Academy of Sciences* 930:43–61.

Huyse, T., R. Poulin, and A. Théron. 2005. Speciation in parasites: a population genetics approach. *Trends in Parasitology* 21:469–475.

International Panel on Climate Change. 2007. Summary for policymakers. In *Climate Change 2007: The Physical Science Basis. Contribution of Working Group I to the Fourth Assessment Report of the Intergovernmental Panel on Climate Change*, edited by S. Solomon, D. Qin, M. Manning, et al. Cambridge University Press, New York.

Jackson, A. P. 2005. The effect of paralogous lineages on the application of reconciliation analysis by cophylogeny mapping. *Systematic Biology* 54:127–145.

Jackson, A. P., and M. A. Charleston. 2004. A cophylogenetic perspective of RNA-virus evolution. *Molecular Biology and Evolution* 21:45–57.

Jaenike, J. 1990. Host specialization in phytophagous insects. *Annual Review of Ecology and Systematics* 21:243–273.

James, P. J. 1999. Do sheep regulate the size of their mallophagan louse populations? *International Journal for Parasitology* 29:869–875.

James, P. J., I. H. C. Carmichael, A. Pfeffer, R. R. Martin, and M. G. O'Callaghan. 2002. Variation among Merino sheep in susceptibilty to lice (*Bovicola ovis*) and association with susceptibility to trichostrongylid gastrointestinal parasites. *Veterinary Parasitology* 103:355–365.

James, P. J., and R. D. Moon. 1998. Pruritis and dermal response to insect antigens in sheep infested with *Bovicola ovis*. *International Journal for Parasitology* 28:419–427.

Janz, N. 2011. Ehrlich and Raven revisited: mechanisms underlying codiversification of plants and enemies. *Annual Review of Ecology, Evolution, and Systematics* 42:71–89.

Janzen, D. H. 1968. Host plants as islands in evolutionary and contemporary time. *American Naturalist* 102:592–595.

———. 1973. Host plants as islands. II. Competition in evolutionary and contemporary time. *American Naturalist* 107:786–790.

———. 1980. When is it coevolution? *Evolution* 34:611–612.

———. 1985a. Coevolution as a process: what parasites of animals and plants do not have in common. In *Coevolution of Parasitic Arthropods and Mammals*, edited by K. C. Kim, 83–99. Wiley and Sons, New York.

———. 1985b. Dan Janzen's thoughts from the tropics. 1. On ecological fitting. *Oikos* 45:308–310.

Johnson, K. P., R. J. Adams, and D. H. Clayton. 2002a. The phylogeny of the louse genus *Brueelia* does not reflect host phylogeny. *Biological Journal of the Linnaean Society* 77:233–247.

Johnson, K. P., R. J. Adams, R. D. M. Page, and D. H. Clayton. 2003. When do parasites fail to speciate in response to host speciation? *Systematic Biology* 52:37–47.

Johnson, K. P., J. M. Allen, B. P. Olds, et al. 2014. Rates of genomic divergence in humans, chimpanzees, and their lice. *Proceedings of the Royal Society of London B* 281:20132174.

Johnson, K. P., S. E. Bush, and D. H. Clayton. 2005. Correlated evolution of host and parasite body size: tests of Harrison's rule using birds and lice. *Evolution* 59: 1744–1753.

Johnson, K. P., and D. H. Clayton. 2003a. The biology, ecology, and evolution of chewing lice. In *The Chewing Lice: World Checklist and Biological Overview*, edited by R. D. Price, R. A. Hellenthal, R. L. Palma, K. P. Johnson, and D. H. Clayton, 449–476. Illinois Natural History Survey Special Publication 24.

———. 2003b. Coevolutionary history of ecological replicates: comparing phylogenies of wing and body lice to Columbiform hosts. In *Tangled Trees: Phylogeny, Cospeciation and Coevolution*, edited by R. D. M. Page, 262–286. University of Chicago Press, Chicago.

———. 2004. Untangling coevolutionary history. *Systematic Biology* 53:92–94.

Johnson, K. P., M. Kennedy, and K. G. McCracken. 2006. Reinterpreting the origins of flamingo lice: cospeciation or host-switching? *Biology Letters* 2:275–278.

Johnson, K. P., J. R. Malenke, and D. H. Clayton. 2009. Competition promotes the evolution of host generalists in obligate parasites. *Proceedings of the Royal Society of London B* 276:3921–3926.

Johnson, K. P., D. L. Reed, S. Hammond, D. Kim, and D. H. Clayton. 2007. Phylogenetic analysis of nuclear and mitochondrial genes supports species groups for *Columbicola* (Insecta: Phthiraptera). *Molecular Phylogenetics and Evolution* 45:506–518.

Johnson, K. P., and J. Seger. 2001. Elevated rates of nonsynonymous substitution in island birds. *Molecular Biology and Evolution* 18:874–881.

Johnson, K. P., S. M. Shreve, and V. S. Smith. 2012. Repeated adaptive divergence of microhabitat specialization in avian feather lice. *BMC Biology* 10:52.

Johnson, K. P., J. D. Weckstein, S. E. Bush, and D. H. Clayton. 2011a. The evolution of host specificity in dove body lice. *Parasitology* 138:1730–1736.

Johnson, K. P., J. D. Weckstein, M. J. Meyer, and D. H. Clayton. 2011b. There and back again: switching between host orders by avian body lice (Ischnocera: Goniodidae). *Biological Journal of the Linnean Society* 102:614–625.

Johnson, K. P., B. L. Williams, D. M. Drown, R. J. Adams, and D. H. Clayton. 2002b. The population genetics of host specificity: genetic differentiation in dove lice (Insecta: Phthiraptera). *Molecular Ecology* 11:25–38.

Johnson, K. P., K. Yoshizawa, and V. S. Smith. 2004. Multiple origins of parasitism in lice. *Proceedings of the Royal Society of London B* 271:1771–1776. doi:10.1098/rspb .2004.2798.

Johnson, P. T. J., A. Dobson, K. D. Lafferty, et al. 2010. When parasites become prey: ecological and epidemiological significance of eating parasites. *Trends in Ecology and Evolution* 25:362–371.

Johnston, R. F., and M. Janiga. 1995. *Feral Pigeons*. Vol. 4. Oxford University Press, Oxford.

Jokela, J., M. F. Dybdahl, and C. M. Lively. 2009. The maintenance of sex, clonal dynamics, and host-parasite coevolution in a mixed population of sexual and asexual snails. *American Naturalist* 174: S43–53.

Jones, C. J. 1996. Immune responses to fleas, bugs and sucking lice. In *The Immunology of Host-Ectoparasitic Arthropod Relationships*, edited by S. K. Wikel, 150–173. CAB International, Wallingford, UK.

Jones, D. 1998. The neglected saliva: medically important toxins in the saliva of human lice. *Parasitology* 116:S73–82.

Jones, I. L. 1993. Crested auklet (*Aethia cristatella*). In *The Birds of North America Online*, edited by A. Poole. Ithaca: Cornell Lab of Ornithology. doi:10.2173/bna.70.

Jousselin, E., Y. Desdevises, and A. Coeur d'Acier. 2009. Fine-scale cospeciation between *Brachycaudus* and *Buchnera aphidicola*: bacterial genome helps define species and evolutionary relationships in aphids. *Proceedings of the Royal Society of London B* 276:187–196.

Kamyszek, F., and W. Gibasiewicz. 1986. Pediculosis in pigs in the light of clinical and laboratory studies. *Wiadomości Parazytologiczne* 32:191–197.

Karvonen, A., and O. Seehausen. 2012. The role of parasitism in adaptive radiations— When might parasites promote and when might they constrain ecological speciation? *International Journal of Ecology* doi:10.1155/2012/280169.

Kassen, R. 2009. Toward a general theory of adaptive radiation: insights from microbial experimental evolution. *Annals of the New York Academy of Science* 1168: 3–22.

Kavaliers, M., E. Choleris, and D. W. Pfaff. 2005. Recognition and avoidance of the odors of parasitized conspecifics and predators: differential genomic correlates. *Neuroscience and Biobehavioral Reviews* 29:1347–1359.

Kavaliers, M., D. D. Colwell, E. Choleris, et al. 2003a. Impaired discrimination of and aversion to parasitized male odors by female oxytocin knockout mice. *Genes, Brain and Behavior* 2:220–230.

Kavaliers, M., M. A. Fudge, D. D. Colwell, and E. Choleris. 2003b. Aversive and avoidance responses of female mice to the odors of males infected with an ectoparasite and the effects of prior familiarity. *Behavioral Ecology and Sociobiology* 54:423–430.

Keirans, J. E. 1975a. A review of the phoretic relationship between Mallophaga (Phthiraptera: Insecta) and the Hippoboscidae (Diptera: Insecta). *Journal of Medical Entomology* 12:71–76.

———. 1975b. Records of phoretic attachment of Mallophaga (Insecta: Phthiraptera) on insects other than Hippoboscidae. *Journal of Medical Entomology* 12:476.

Kéler, V. S. 1952. Uber den feineren Bau der Tarsen bei *Pseudomenopon rowanae* Kéler. *Beiträge zur Entomologie* 2:573–582.

Kethley, J. B., and D. E. Johnston. 1975. Resource tracking patterns in bird and mammal ectoparasites. *Miscellaneous Publications of the Entomological Society of America* 9:231–236.

Khater, H. F., M. Y. Ramadan, and R. S. El-Madawy. 2009. Lousicidal, ovicidal and repellent efficacy of some essential oils against lice and flies infesting water buffaloes in Egypt. *Veterinary Parasitology* 164:257–266.

Kiester, A. R., R. Lande, and D. W. Schemske. 1984. Models of coevolution and speciation in plants and their pollinators. *American Naturalist* 124:220–243.

Kim, D. 2008. *Evolution of crypsis in feather lice*. Master's thesis, University of Utah.

Kim, K. C. 1972. Louse populations of the northern fur seal (*Callorhinus ursinus*). *American Journal of Veterinary Research* 33:2027–2036.

———. 1975. Ecology and morphological adaptation of the sucking lice (Anoplura, Echinophthiriidae) on the northern fur seal. *Rapport et Procès verbaux des Réunions du conseil Permanent International pour l'Exploration de la Mer* 169:504–514.

———. 1985a. Evolution and host associations of Anoplura. In *Coevolution of Parasitic Arthropods and Mammals*, edited by K. C. Kim, 197–231. John Wiley and Sons, New York.

———. 1985b. Evolutionary relationships of parasitic arthropods and mammals. In *Coevolution of Parasitic Arthropods and Mammals*, edited by K. C. Kim, 3–82. John Wiley and Sons, New York.

Kim, K. C., K. C. Emerson, and R. D. Price. 1973. Lice. In *Parasites of Laboratory Animals*, edited by R. J. Flynn, 376–397. Iowa State University Press, Ames.

Kingsolver, J. G., H. E. Hoekstra, J. M. Hoekstra, et al. 2001. The strength of phenotypic selection in natural populations. *American Naturalist* 157:245–261.

Kirk, W. D. 1991. The size relationship between insects and their hosts. *Ecological Entomology* 16:351–359.

Kirkness, E. F., B. J. Haas, W. Sun, et al. 2010. Genome sequences of the human body louse and its primary endosymbiont provide insights into the permanent parasitic lifestyle. *Proceedings of the National Academy of Sciences USA* 107:12168–12173.

Kirk-Spriggs, A. H., and E. Mey. 2014. Phoresy of a sucking louse, *Linognathus* sp. (Phthiraptera: Anoplura: Linognathidae), by *Musca (Byomya) conducens* Walker (Diptera: Muscidae) in South Africa. *African Invertebrates* 55:119–123.

Kittler, R., M. Kayser, and M. Stoneking. 2003. Molecular evolution of *Pediculus humanus* and the origin of clothing. *Current Biology* 13:1414–1417.

Koh, L., R. Dunn, N. Sodhi, R. Colwell, H. Proctor, and V. Smith. 2004. Species coextinctions and the biodiversity crisis. *Science* 305:1632.

Koop, J. A. H., K. E. DeMatteo, P. G. Parker, and N. K. Whiteman 2014. Birds are islands for parasites. *Biology Letters* 10:20140255.

Kose, M., R. Mand, and A. P. Møller. 1999. Sexual selection for white tail spots in the barn swallow in relation to habitat choice by feather lice. *Animal Behavior* 58: 1201–1205.

Kounek, F., O. Sychra, M. Capek, and I. Literak. 2011. Chewing lice of the genus *Myrsidea* (Phthiraptera: Menoponidae) from New World warblers (Passeriformes: Parulidae) from Costa Rica, with descriptions of four new species. *Zootaxa* 3137: 56–63.

Krasnov, B. R. 2008. *Functional and Evolutionary Ecology of Fleas*. Cambridge University Press, Cambridge.

Krasnov, B. R., N. V. Burdelova, I. S. Khokhlova, G. I. Shenbrot, and A. Degen. 2005a. Larval interspecific competition in two flea species parasitic on the same rodent host. *Ecological Entomology* 30:146–155.

Krasnov, B. R., D. Mouillot, G. I. Shenbrot, I. S. Khokhlova, and R. Poulin. 2005b. Abundance patterns and coexistence processes in communities of fleas parasitic on small mammals. *Ecography* 28:453–464.

Krebs, C. J. 2009. *Ecology*. 6th ed. Benjamin Cummings, San Francisco.

Kuris, A. M. 2012. The global burden of human parasites: who and where are they? How are they transmitted? *Journal of Parasitology* 98:1056–1064.

Kuris, A. M., A. R. Blaustein, and J. Javier Alió. 1980. Hosts as islands. *American Naturalist* 116:570–586.

Kuris, A. M., and K. D. Lafferty. 1994. Community structure: larval trematodes in snail hosts. *Annual Review of Ecology and Systematics* 25:189–217.

Kwon, D. H., K. S. Yoon, J. P. Strycharz, J. M. Clark, and S. H. Lee. 2008. Determination of permethrin resistance allele frequency of human head louse populations by quantitative sequencing. *Journal of Medical Entomology* 45: 912–920.

Labandeira, C. C. 2013. A paleobiologic perspective on plant-insect interactions. *Current Opinion in Plant Biology* 16:414–421.

Labandeira, C. C., D. L. Dilcher, D. R. Davis, and D. L. Wagner. 1994. Ninety-seven million years of angiosperm-insect association: paleobiological insights into the meaning of coevolution. *Proceedings of the National Academy of Sciences USA* 91: 12278–12282.

Labandeira, C. C., S. L. Tremblay, K. E. Bartowski, and L. V. Hernick. 2014. Middle

Devonian liverwort herbivory and antiherbivore defence. *New Phytologist* 202: 247–258.

Lafferty, K. D. 2010. Interacting parasites. *Science* 330:187–188.

Lans C., and N. Turner. 2011. Organic parasite control for poultry and rabbits in British Columbia, Canada. *Journal of Ethnobiology and Ethnomedicine* 7:21. doi: 10.1186/1746-4269-7-21.

Lapoint, R., and N. Whiteman. 2012. How a bird is an island. *BMC Biology* 10:53.

Larson, G. 1989. PreHistory of The Far Side®: A 10th Anniversary Exhibit. Andrews McMeel Publishing. Kansas City.

Laukkanen, L., R. Leimu, A. Muola, M. Lilley, J. P. Salminen, and P. Mutikainen. 2012. Plant chemistry and local adaptation of a specialized folivore. *PLOS One* 7:e38225.

Lee, P. L. M., and D. H. Clayton. 1995. Population biology of swift (*Apus apus*) ectoparasites in relation to host reproductive success. *Ecological Entomology* 20: 43–50.

Lee, P. L. M., D. Clayton, R. Griffiths, and R. Page. 1996. Does behavior reflect phylogeny in swiftlets (Aves: Apodidae)? A test using cytochrome b mitochondrial DNA sequences. *Proceedings of the National Academy of Sciences* 93:7091.

Lehane, M. J. 2005. *The Biology of Blood-sucking Insects*. 2nd ed. Cambridge University Press, Cambridge.

Lehmann, T. 1993. Ectoparasites: direct impact on host fitness. *Parasitology Today* 9: 8–13.

Leigh, E. G., Jr. 1970. Natural selection and mutability. *American Naturalist* 104: 301–305.

Lello, J., B. Boag, A. Fenton, I. R. Stevenson, and P. J. Hudson. 2004. Competition and mutualism among the gut helminths of a mammalian host. *Nature* 428:837–840.

Lemaire, B., S. Huysmans, E. Smets, and V. Merckx. 2011. Rate accelerations in nuclear 18S rDNA of mycohetertrophic and parasitic angiosperms. *Journal of Plant Research* 5:561–576.

Leo, N. P., J. M. Hughes, X. Yang, S. K. S. Poudel, W. G. Brogdon, and S. C. Barker. 2005. The head and body lice of humans are genetically distinct (Insecta: Phthiraptera: Pediculidae): evidence from double infestations. *Heredity* 95:34–40.

Leonardi, M. S., E. A. Crespo, J. A. Raga, and F. J. Aznar. 2013. Lousy mums: patterns of vertical transmission of an amphibious louse. *Parasitology Research* 112:3315–3323.

Leonardi, M. S., E. A. Crespo, J. A. Raga, and M. Fernández. 2012. Scanning electron microscopy of *Antarctophthirus microchir* (Phthiraptera: Anoplura: Echinophthiriidae): studying morphological adaptations to aquatic life. *Micron* 43:929–936.

Levin, S. A. 2009. *The Princeton Guide to Ecology*. Princeton University Press, Princeton.

Levins, R. 1969. Some demographic and genetic consequences of environmental heterogeneity for biological control. *Bulletin of the Entomological Society of America* 15:237–240.

Levins, R., and D. Culver. 1971. Regional coexistence of species and competition between rare species. *Proceedings of the National Academy of Sciences USA* 68: 1246–1248.

Levot, G. 2000. Resistance and the control of lice on humans and production animals. *International Journal for Parasitology* 30:291–297.

Lewis, L. F., D. M. Christenson, and G. W. Eddy. 1967. Rearing the long-nosed cattle louse and cattle biting louse on host animals in Oregon. *Journal of Economic Entomology* 60:755–757.

Liechti, F., W. Witvliet, R. Weber, and E. Bächler. 2013. First evidence of a 200-day non-stop flight in a bird. *Nature Communications* 4:1–7.

Light, J. E., and M. S. Hafner. 2007. Cophylogeny and disparate rates of evolution in sympatric lineages of chewing lice on pocket gophers. *Molecular Phylogenetics and Evolution* 45:997–1013.

———. 2008. Codivergence in heteromyid rodents (Rodentia: Heteromyidae) and their sucking lice of the genus *Fahrenholzia* (Phthiraptera: Anoplura). *Systematic Biology* 57:449–465.

Light, J. E., V. S. Smith, J. M. Allen, L. A. Durden, and D. L. Reed. 2010. Evolutionary history of mammalian sucking lice (Phthiraptera:Anoplura). *BMC Evolutionary Biology* 10:292. http://www.biomedcentral.com/1471-2148/10/292.

Light, J. E., M. A. Toups, and D. L. Reed. 2008. What's in a name: the taxonomic status of human head and body lice. *Molecular Phylogenetics and Evolution* 47: 1203–1216.

Linsdale. J. M., and L. P. Tevis Jr. 1951. *The Dusky-Footed Wood Rat*. University of California Press, Berkeley.

Lion, S., M. van Baalen, and W. G. Wilson. 2006. The evolution of parasite manipulation of host dispersal. *Proceedings of the Royal Society of London B* 273: 1063–1071.

Lodmell, D. L., J. F. Bell, C. M. Clifford, G. J. Moore, and G. Raymond. 1970. Effects of limb disability on lousiness of mice. V. Hierarchy disturbance on mutual grooming and reproductive capacities. *Experimental Parasitology* 27:184–192.

Loehle, C. 1997. The pathogen transmission avoidance theory of sexual selection. *Ecological Modelling* 103:231–250.

Loker, E. S. 2012. Macroevolutionary immunology: a role for immunity in the diversification of animal life. *Frontiers in Immunology* 3:25. doi:10.3389/fimmu.2012.00025.

Longino, J. T. 1984. True anting by the capuchin, *Cebus capucinus*. *Primates* 25: 243–245.

Loomis, E. C. 1978. External parasites. In *Diseases of Poultry*, 7th ed., edited by M. S. Hofstad, 667–704. Iowa State University Press, Ames.

Lopez-Pascua, L. L., S. Gandon, and A. Buckling. 2012. Abiotic heterogeneity drives parasite local adaptation in coevolving bacteria and phages. *Journal of Evolutionary Biology* 25:187–195.

Losos, J. B. 2009. *Lizards in an Evolutionary Tree: Ecology and Adaptive Radiation of Anoles*. Vol. 10. University of California Press, Berkeley.

Losos, J. B., T. R. Jackman, A. Larson, K. De Queiroz, and L. Rodriguez-Schettino.

1998. Contingency and determinism in replicated adaptive radiations of island lizards. *Science* 279:2115–2118.

Luong, L. T., B. D. Heath, and M. Polak. 2007. Host inbreeding increases susceptibility to ectoparasitism. *Journal of Evolutionary Biology* 20:79–86.

Lyal, C. H. 1985. Phylogeny and classification of the Psocodea, with particular reference to the lice (Psocodea: Phthiraptera). *Systematic Entomology* 10:145–165.

Lyotard, J. F. 1984. *The Postmodern Condition: A Report on Knowledge*. Originally published 1979. Manchester University Press, Manchester.

Maa, T. C. 1969. Notes on the Hippoboscidae (Diptera). *Pacific Insects* 4:583–614.

MacColl, A. D. C., and S. M. Chapman. 2010. Parasites can cause selection against migrants following dispersal between environments. *Functional Ecology* 24:847–856.

MacLeod, C. J., A. M. Paterson, D. M. Tompkins, and R. P. Duncan. 2010. Parasites lost—do invaders miss the boat or drown on arrival? *Ecology Letters* 13:516–527.

Maddison, W. P. 1990. A method for testing the correlated evolution of two binary characters: are gains or losses concentrated on certain branches of a phylogenetic tree? *Evolution* 44:539–557.

Maestripieri, D. 1993. Vigilance costs of allogrooming in macaque mothers. *American Naturalist* 141:744–753.

Maiolino, S., D. M. Boyer, and A. Rosenberger. 2011. Morphological correlates of the grooming claw in distal phalanges of Platyrrhines and other primates: a preliminary study. *Anatomical Record* 294:1975–1990.

Majerus, M. 1998. *Melanism: Evolution in Action*. Oxford University Press, Oxford.

Malenke, J. R., K. P. Johnson, and D. H. Clayton. 2009. Host specialization differentiates cryptic species of feather-feeding lice. *Evolution* 63:1427–1438.

Malenke, J. R., N. Newbold, and D. H. Clayton. 2011. Condition-specific competition governs the geographic distribution and diversity of ectoparasites. *American Naturalist* 177:522–534.

Malthus, T. P. 1798. *An Essay on the Principle of Population*. 1998 reprint. Prometheus Books, Amherst.

Manjerovic, M. B., and J. M. Waterman. 2012. Immunological sex differences in socially promiscuous African ground squirrels. *PLoS ONE* 7:e38524.

Marshall, A. G. 1981a. *The Ecology of Ectoparasitic Insects*. Academic Press, London.

———. 1981b. The sex ratio in ectoparasitic insects. *Ecological Entomology* 6:155–174.

———. 1982. Ecology of insects ectoparasitic on bats. In *Ecology of Bats*, edited by T. H. Kunz, 369–401. Plenum Press, New York.

Martin, C. D., and B. A. Mullens. 2012. Housing and dustbathing effects on northern fowl mites (*Ornithonyssus sylviarum*) and chicken body lice (*Menacanthus stramineus*) on hens. *Medical and Veterinary Entomology* 26:323–333.

Martinez-Padilla, J., P. Vergara, F. Mougeot, and S. M. Redpath. 2012. Parasitized mates increase infection risk for partners. *American Naturalist* 179:811–820.

Matthee, S., I. G. Horak, and D. G. A. Meltzer. 1998. The distribution and seasonal changes of louse populations on impala (*Aepyceros melampus*). *South African Journal of Wildlife Research* 28:22–25.

Matthysse, J. G. 1946. Cattle lice: their biology and control. *Bulletin Cornell University Agricultural Experiment Station* 832:1–67.

May, R. M., and R. M. Anderson. 1978. Regulation and stability of host-parasite population interactions. II. Destabilizing processes. *Journal of Animal Ecology* 47: 249–267.

———. 1983. Epidemiology and genetics in the coevolution of parasites and hosts. *Proceedings of the Royal Society of London B* 219:281–313.

Mayr, E. 1963. *Animal Species and Evolution.* Harvard University Press, Cambridge.

McCallum, H., and A. Dobson. 1995. Detecting disease and parasite threats to endangered species and ecosystems. *Trends in Ecology and Evolution* 10:190–194.

McClearn, D. 1992. The rise and fall of a mutualism? Coatis, tapirs, and ticks on Barro Colorado Island, Panamá. *Biotropica* 24:220–222.

McCoy, K. D., T. Boulinier, C. Tirard, and Y. Michalakis. 2003. Host-dependent genetic structure of parasite populations: differential dispersal of seabird tick host races. *Evolution* 57:288–296.

McGraw, K. J. 2006. Mechanics of melanin-based coloration. In *Bird Coloration, Vol. 1: Mechanisms and Measurements*, edited by G. E. Hill and K. J. McGraw, 243–294. Harvard University Press, Cambridge.

McKenzie, A. A. 1990. The ruminant dental grooming apparatus. *Zoological Journal of the Linnean Society* 99:117–128.

McKenzie, A. A., and A. Weber. 1993. Loose front teeth: radiological and histological correlation with grooming function in the impala *Aepyceros melampus*. *Journal of Zoology*. 231:167–174.

Medzhitov, R., D. S. Schneider, and M. P. Soares. 2012. Disease tolerance as a defense strategy. *Science* 335:936–941.

Merilä, J., and D. Wiggin. 1995. Interspecific competition for nest holes causes adult mortality in the collared flycatcher. *Condor* 97:445–450.

Mey, E. 2013. A phenomenon noted particularly in birds: drinking of eye secretion (lachrymophagy) by lice (Insecta, Phthiraptera). *Vogelwarte* 51:15–23.

Meyer, A., T. D. Kocher, P. Basasibawaki, and A. C. Wilson. 1990. Monophyletic origin of Lake Victoria cichlid fishes. *Nature* 347:550–553.

Mideo, N. 2009. Parasite adaptations to within-host competition. *Trends in Parasitology* 25:261–268.

Mitter, C., B. Farrell, and B. Wiegmann. 1988. The phylogenetic study of adaptive zones: has phytophagy promoted insect diversification? *American Naturalist* 132: 107–128.

Mitzmain, M. B. 1912. Collected notes on the insect transmission of surra in carabaos. *Philippine Agricultural Review* 5:670–681.

Mode, C. J. 1958. A mathematical model for the co-evolution of obligate parasites and their hosts. *Evolution* 12:158–165.

Møller, A. P. 1991. Parasites, sexual ornaments and mate choice in the barn swallow *Hirundo rustica*. In *Bird-Parasite Interactions: Ecology, Evolution, and Behaviour*, edited by J. E. Loye and M. Zuk, 328–343. Oxford University Press, Oxford.

Møller, A. P., P. Christe, and E. Lux. 1999. Parasitism, host immune function, and sexual selection. *Quarterly Review of Biology* 74:3–20.

Møller, A. P., F. de Lope, and N. Saino. 2004a. Parasitism, immunity, and arrival date in a migratory bird, the barn swallow. *Ecology* 85:206–219.

Møller, A. P., J. Erritzoe, and L. Rózsa. 2010. Ectoparasites, uropygial glands and hatching success in birds. *Oecologia* 163:303–311.

Møller, A. P., R. Martinelli, and N. Saino. 2004b. Genetic variation in infestation with a directly transmitted ectoparasite. *Journal of Evolutionary Biology* 17:41–47.

Møller, A. P., and L. Rózsa. 2005. Parasite biodiversity and host defenses: chewing lice and immune response of their avian hosts. *Oecologia* 142:169–176.

Moore, J. 2002. *Parasites and the Behavior of Animals.* Oxford University Press, Oxford.

Mooring, M. S. 1995. The effect of tick challenge on grooming rate by impala. *Animal Behavior* 50:377–392.

Mooring, M. S., J. E. Benjamin, C. R. Harte, and N. B. Herzog. 2000. Testing the interspecific body size principle in ungulates: the smaller they come, the harder they groom. *Animal Behavior* 60:35–45.

Mooring, M. S., D. T. Blumstein, and C. J. Stoner. 2004. The evolution of parasite-defence grooming in ungulates. *Biological Journal of the Linnean Society* 81:17–37.

Mooring, M. S., and B. L. Hart. 1995. Costs of allogrooming in impala: distraction from vigilance. *Animal Behavior* 49:1414–1416.

Mooring, M. S., A. A. McKenzie, and B. L. Hart. 1996. Grooming in impala: role of oral grooming in removal of ticks and effects of ticks in increasing grooming rate. *Physiology and Behavior* 59:965–971.

Mooring, M. S., and W. M. Samuel. 1999. Premature loss of winter hair in free-ranging moose (*Alces alces*) infested with winter ticks (*Dermacentor albipictus*) is correlated with grooming rate. *Canadian Journal of Zoology* 77:148–156.

Morand, S., M. S. Hafner, R. D. M. Page, and D. L. Reed. 2000. Comparative body size relationships in pocket gophers and their chewing lice. *Biological Journal of the Linnean Society* 70:239–249.

Moreau, R. E. 1933. The food of the red-billed oxpecker, *Buphagus erythrorhynchus* (Stanley). *Bulletin of Entomological Research* 24:325–335.

Moreno-Rueda, G. 2005. Is the white wing-stripe of male house sparrows (*Passer domesticus*) an indicator of the load of Mallophaga? *Ardea* 93:109–114.

Moreno-Rueda, G. 2010. Uropygial gland size correlates with feather holes, body condition and wingbar size in the house sparrow *Passer domesticus. Journal of Avian Biology* 41:229–236.

Moreno-Rueda, G. 2014. Uropygial gland size, feather holes and moult performance in the house sparrow *Passer domesticus. Ibis* 156:457–460.

Moreno-Rueda, G., and H. Hoi. 2011. Female house sparrows prefer big males with a large white wing bar and fewer feather holes caused by chewing lice. *Behavioral Ecology* 23:271–277.

Morgan, A. D., S. Gandon, and A. Buckling. 2005. The effect of migration on local adaptation in a coevolving host–parasite system. *Nature* 437:253–256.

Morris, R. J., O. T. Lewis, and H. C. J. Godfray. 2004. Experimental evidence for apparent competition in a tropical forest food web. *Nature* 428:310–313.

Moulia, C., J. P. Aussel, F. Bonhomme, P. Boursot, J. T. Nielsen, and F. Renaud. 1991. Wormy mice in a hybrid zone: a genetic control of susceptibility to parasite infection. *Journal of Evolutionary Biology* 4:679–687.

Mountainspring, S., and J. M. Scott. 1985. Interspecific competition among Hawaiian forest birds. *Ecological Monographs* 55:219–239.

Mouritsen, K. N., and J. Madsen. 1994. Toxic birds: defence against parasites? *Oikos* 69:357–358.

Mousseau, T. A., and D. A. Roff. 1987. Natural selection and the heritability of fitness components. *Heredity* 59:181–197.

Moyer, B. R., and D. H. Clayton. 2004. Avian defenses against ectoparasites. In *Insect and Bird Interactions*, edited by H. F. van Emden and M. Rothschild, 241–257. Intercept, Andover, UK.

Moyer, B. R., D. Drown, and D. Clayton. 2002a. Low humidity reduces ectoparasite pressure: implications for host life history evolution. *Oikos* 97:223–228.

Moyer, B. R., D. W. Gardiner, and D. H. Clayton. 2002b. Impact of feather molt on ectoparasites: looks can be deceiving. *Oecologia* 131:203–210.

Moyer, B. R., A. T. Peterson, and D. H. Clayton. 2002c. Influence of bill shape on ectoparasite load in Western scrub-jays. *Condor* 104:675–678.

Moyer, B. R., A. N. Rock, and D. H. Clayton. 2003. An experimental test of the importance of preen oil in rock doves (*Columba livia*). *Auk* 120:490–496.

Moyer, B. R., and G. E. Wagenbach. 1995. Sunning by black noddies (*Anous minutus*) may kill chewing lice (*Quadraceps hopkinsi*). *Auk* 112:1073–1077.

Mullen, G. R., and Durden, L. A., eds. 2009. *Medical and Veterinary Entomology*. 2nd ed. Academic Press/Elsevier Science, San Diego.

Muller, H. J. 1964. The relation of recombination to mutational advance. *Mutation Research* 1:2–9.

Mumcuoglu, K. Y. 2008a. Human lice: Pediculus and Pthirus. In *Paleomicrobiology: Past Human Infections*, edited by D. Raoult and M. Drancourt, 215–222. Springer-Verlag, Berlin.

———. 2008b. The louse comb: past and present. *American Entomologist* 54:164–166.

Mumcuoglu, K. Y., D. Ben-Yakir, J. O. Ochanda, J. Miller, and R. Galun. 1997. Immunization of rabbits with faecal extract of *Pediculus humanus*, the human body louse: effects on louse development and reproduction. *Medical and Veterinary Entomology* 11:315–318.

Mumcuoglu, K. Y., R. Galun, Y. Kaminchik, A. Panet, and A. Levanon. 1996. Antihemostatic activity in salivary glands of the human body louse, *Pediculus humanus humanus* (Anoplura: Pediculidae). *Journal of Insect Physiology* 11:1083–1087.

Mumcuoglu, K. Y., S. Magdassi, J. Miller, et al. 2004. Repellency of citronella for head lice: double-blind randomized trial of efficacy and safety. *Israel Medical Association Journal* 6:756–759.

Murray, A. 1861. On the *Pediculi* infesting the different races of man. *Transactions of the Royal Society of Edinburgh* 22:567–578.

Murray, M. D. 1957. The distribution of the eggs of mammalian lice on their hosts. IV. The distribution of the eggs of *Damalinia equi* (Denny) and *Haematopinus asini* (L.) on the horse. *Australian Journal of Zoology* 5:183–187.

———. 1961. The ecology of the louse *Polyplax serrata* (Burm.) on the mouse *Mus musculus* L. *Australian Journal of Zoology* 9:1–13.

———. 1963. The ecology of lice on sheep. IV. The establishment and maintenance of populations of *Linognathus ovillus* (Neumann). *Australian Journal of Zoology* 11: 157–172.

———. 1968. Ecology of lice on sheep. VI. The influence of shearing and solar radiation on populations and transmission of *Damalinia ovis*. *Australian Journal of Zoology* 16:725–738.

———. 1987. Effects of host grooming on louse populations. *Parasitology Today* 3: 276–278.

———. 1990. Influence of host behaviour on some ectoparasites of birds and mammals. In *Parasitism and Host Behaviour*, edited by C. J. Barnard and J. M. Behnke, 290–315. Taylor and Francis, London.

Murray, M. D., and J. H. Calaby. 1971. The host relations of the Boopiidae. *Australian Journal of Zoology (Supplementary Series)* 6:81–84.

Murray, M. D., and D. G. Nicholls. 1965. Studies on the ectoparasites of seals and penguins. I. The ecology of the louse *Lepidophthirus macrorhini* Enderlein on the southern elephant seal, *Mirounga leonina* (L.). *Australian Journal of Zoology* 13: 437–454.

Murray, M. D., M. S. R. Smith, and Z. Soucek. 1965. Studies on the ectoparasites of seals and penguins. 2. The ecology of the louse *Antarctophthirus ogmorhini* Enderlein on the Weddell seal, *Leptonychotes weddelli* Lesson. *Australian Journal of Zoology* 13:761–772.

Murrell, A., and S. C. Barker. 2005. Multiple origins of parasitism in lice:phylogenetic analysis of SSU rDNA indicates that Phthiraptera and Psocoptera are not monophyletic. *Parasitology Research* 97:274–280. doi:10.1007/s00436-005-1413-8.

Nabokov, V. 1963. *The Gift*. Translated by M. Scammell in collaboration with V. Nabokov. Putnam, New York.

Nadler, S. A. 1995. Microevolution and the genetic structure of parasite populations. *Journal of Parasitology* 81:395–403.

Nadler, S. A., M. S. Hafner, J. C. Hafner, and D. J. Hafner. 1990. Genetic differentiation among chewing louse populations (Mallophaga: Trichodectidae) in a pocket gopher contact zone (Rodentia: Geomyidae). *Evolution* 44:942–951.

Najer, T., O. Sychra, N. M. Hung, M. Capek, P. Podzemny, and I. Literak. 2012. Chewing lice (Phthiraptera: Amblycera, Ischnocera) from wild passerines (Aves: Passeriformes) in northern Vietnam, with descriptions of three new species. *Zootaxa* 3530:59–73.

Nash, D. R. 2008. Process rather than pattern: finding pine needles in the coevolutionary haystack. *Journal of Biology* 7:14.

Nelson, B. C., and M. D. Murray. 1971. The distribution of Mallophaga on the domestic pigeon (*Columba livia*). *International Journal for Parasitology* 1:21–22.

Nelson, H., and G. Geher. 2007. Mutual grooming in human dyadic relationships:an ethological perspective. *Current Psychology* 26:121–140.

Nelson, W. A. 1984. Effects of nutrition of animals on their ectoparasites. *Journal of Medical Entomology* 21:621–635.

———. 1987. Other blood-sucking and myiasis-producing arthropods: immune responses in parasitic infections: immunology, immunopathology and immunoprophylaxis. In *Protozoa, Arthropods and Invertebrates, Vol. IV*, edited by E. J. L. Soulsby, 175–209. CRC, Boca Raton.

Nelson, W. A., J. F. Bell, C. M. Clifford, and J. E. Keirans. 1977. Interaction of ectoparasites and their hosts. *Journal of Medical Entomology* 13:389–428.

Nelson, W. A., J. F. Bell, and S. J. Stewart. 1979. *Polyplax serrata*: cutaneous cytologic reactions in mice that do (CFW strain) and do not (C57BL) develop resistance. *Experimental Parasitology* 48:259–264.

Nelson, W. A., J. F. Bell, S. J. Stewart, and G. C. Kozub. 1983. *Polyplax serrata* (Phthiraptera:Hoplopleuridae):effects of cortisone and cyclophosphamide on the acquired resistance response of mice to lice. *Journal of Medical Entomology* 20: 551–557.

Nelson, W. A., J. A. Shemanchuk, and W. O. Haufe. 1970. *Haematopinus eurysfernus*: blood of cattle infested with the short-nosed cattle louse. *Experimental Parasitology* 28:263–271.

Nieberding, C., S. Morand, R. Libois, and J. R. Michaux. 2004. A parasite reveals cryptic phylogeographic history of its host. *Proceedings of the Royal Society of London B* 271:2559–2568.

Nieberding, C. M., and I. Olivieri. 2007. Parasites: proxies for host genealogy and ecology? *Trends in Ecology and Evolution* 22:156–165.

NOAA. 2002. Climate atlas of the United States. Version 2. National Climate Data Center, National Oceanic and Atmospheric Administration, US Department of Commerce, Asheville, NC.

Nordberg, S. 1936. Biologisch-ökologische Untersuchungen über die Vogelnidicolen. *Acta Zoologica Fennica* 21:1–168.

Norris, A., K. L. Cockle, and K. Martin. 2010. Evidence for tolerance of parasitism in a tropical cavity-nesting bird, planalto woodcreeper (*Dendrocolaptes playrostris*) in northern Argentina. *Journal Tropical Ecology* 26:619–626.

Nosil, P. 2002. Transition rates between specialization and generalization in phytophagous insects. *Evolution* 56:1701–1706.

———. 2012. *Ecological Speciation*. Oxford University Press, Oxford.

Nosil, P., and A. Mooers. 2005. Testing hypotheses about ecological specialization using phylogenetic trees. *Evolution* 59:2256–2263.

Nuismer, S. L., R. Gomulkiewicz, and B. J. Ridenhour. 2010. When is correlation coevolution? *American Naturalist* 175:525–537.

Núñez-Farfán, J., J. Fornoni, and P. L. Valverde. 2007. The evolution of resistance and tolerance to herbivores. *Annual Review of Ecology and Systematics* 38:541–566.

Nuttall, G. H. 1917. The biology of *Pediculus humanus*. *Parasitology* 10:80–185.

———. 1919a. The biology of *Pediculus humanus*: supplementary notes. *Parasitology* 11:201–220.

———. 1919b. The systematic position, synonymy and iconography of *Pediculus humanus* and *Phthirus pubis*. *Parasitology* 11:329–346.

Nyman, T. 2010. To speciate, or not to speciate? Resource heterogeneity, the subjectivity of similarity, and the macroevolutionary consequences of niche width shifts in plant feeding insects. *Biological Reviews* 85:393–411.

Nyman, T., V. Vikberg, D. R. Smith, and J. L. Boeve. 2010. How common is ecological speciation in plant-feeding insects? A "higher" Nematinae perspective. *BMC Evolutionary Biology* 10:266.

Oguge, N. O. N., L. A. L. Durden, J. E. J. Keirans, H. D. H. Balami, and T. G. T. Schwan. 2009. Ectoparasites (sucking lice, fleas and ticks) of small mammals in southeastern Kenya. *Medical and Veterinary Entomology* 23:387–392.

Ohta, T. 1992. The nearly neutral theory of molecular evolution. *Annual Review of Ecology and Systematics* 23:263–286.

Olds, B. P., B. S. Coates, L. D. Steele, et al. 2012. Comparison of the transcriptional profiles of head and body lice. *Insect Molecular Biology* 21:257–268.

Olson, S. L., and A. Feduccia. 1980. Relationships and evolution of flamingos (Aves: Phoenicopteridae) and grebes (Podicipedidae). *Zoological Journal of the Linnean Society* 140:157–169.

Orr, H. A. 1995. The population genetics of speciation: the evolution of hybrid incompatibilities. *Genetics* 139:1805–1813.

Ortego, J., J. M. Aparicioa, G. Calabuig, and P. J. Cordero. 2007. Risk of ectoparasitism and genetic diversity in a wild lesser kestrel population. *Molecular Ecology* 16: 3712–3720.

Orwell, G. 1938. *Homage to Catalonia.* Secker and Warburg, London.

Otter, A., D. F. Twomey, T. R. Crawshaw, and P. Bates. 2003. Anaemia and mortality in calves infested with the long-nosed sucking louse (*Linognathus vituli*). *Veterinary Record* 153:176–179.

Overington, S. E., J. Morand-Ferron, N. J. Boogert, and L. Lefebvre. 2009. Technical innovations drive the relationship between innovativeness and residual brain size in birds. *Animal Behavior* 78:1001–1010.

Overstreet, R. M. 2003. Flavor buds and other delights. *Journal of Parasitology* 89: 1093–1107.

Owen, J. P., and D. H. Clayton. 2007. Where are the parasites in the PHA response? *Trends in Ecology and Evolution* 22:228–229.

Owen, J. P., M. E. Delany, C. J. Cardona, A. A. Bickford, and B. A. Mullens. 2009. Host inflammatory response governs fitness in an avian ectoparasite, the northern fowl mite (*Ornithonyssus sylviarum*). *International Journal for Parasitology* 39:789–799.

Owen, J. P., A. C. Nelson, and D. H. Clayton. 2010. Ecological immunology of bird-ectoparasite systems. *Trends in Parasitology* 26:530–539.

Page, R. D. M., ed. 2003. *Tangled Trees: Phylogeny, Cospeciation, and Coevolution.* University of Chicago Press, Chicago.

Page, R. D. M. 2006. *Cospeciation*. Wiley Online Library. doi:10.1038/npg. els.0004124.

Page, R. D. M., and M. A. Charleston. 1998. Trees within trees: phylogeny and historical associations. *Trends in Ecology and Evolution* 13:356–359.

Page, R. D. M., R. H. Cruickshank, and K. P. Johnson. 2002. Louse 12S rRNA secondary structure is highly variable. *Insect Molecular Biology* 11:361–369.

Page, R. D. M., R. H. Cruickshank, M. Dickens, et al. 2004. Phylogeny of "*Philoceanus*" complex seabird lice (Phthiraptera: Ischnocera) inferred from mitochondrial DNA sequences. *Molecular Phylogenetics and Evolution* 30:633–652.

Page, R. D. M., P. L. M. Lee, S. A. Becher, R. Griffiths, and D. H. Clayton. 1998. A different tempo of mitochondrial DNA evolution in birds and their parasitic lice. *Molecular Phylogenetics and Evolution* 9:276–293.

Pagel, M. 1994. Detecting correlated evolution on phylogenies: a general method for the comparative analysis of discrete characters. *Proceedings of the Royal Society of London B* 255:37–45.

Pagel, M., and W. Bodmer. 2003. A naked ape would have fewer parasites. *Proceedings of the Royal Society of London B* 270:S117–119.

Paine, R. T. 1966. Food web complexity and species diversity. *American Naturalist* 100: 65–75.

———. 1974. Intertidal community structure. *Oecologia* 15:93–120.

Painter, R. H. 1958. Resistance of plants to insects. *Annual Review of Entomology* 3: 267–290.

Palma, R., and J. Jensen. 2005. Lice (Insecta: Phthiraptera) and their host associations in the Faroe Islands. *Steenstrupia* 29:49–73.

Panagiotakopulu, E., P. C. Buckland, P. M. Day, and C. Doumas. 1995. Natural insecticides and insect repellents in antiquity: a review of the evidence. *Journal of Archaeological Science* 22:705–710.

Paperna, I. 1964. Competitive exclusion of *Dactylogyrus extensus* by *Dactylogyrus vastator* (Trematoda, Monogenea) on the gills of reared carp. *Journal of Parasitology* 50:94–98.

Parchman, T. L., and C. W. Benkman. 2002. Diversifying coevolution between crossbills and black spruce on Newfoundland. *Evolution* 56:1663–1672.

Parsons, P. A. 1996. Competition versus abiotic factors in variably stressful environments: evolutionary implications. *Oikos* 75:129–132.

Paterson, A. M., and J. Banks. 2001. Statistical approaches to measuring cospeciation of host and parasites: through a glass, darkly. *International Journal for Parasitology* 31:1012–1022.

Paterson, A. M., R. L. Palma, and R. D. Gray. 1999. How frequently do avian lice miss the boat? Implications for coevolutionary studies. *Systematic Biology* 48: 214–223.

Paterson, A. M., and R. Poulin. 1999. Have chondracanthid copepods co-speciated with their teleost hosts? *Systematic Parasitology* 44:79–85.

Paterson, S., T. Vogwill, A. Buckling, et al. 2010. Antagonistic coevolution accelerates molecular evolution. *Nature* 464:275–278.

Patton, J. L., M. F. Smith, R. D. Price, and R. A. Hellenthal. 1984. Genetics of hybridization between the pocket gophers *Thomomys bottae* and *Thomomys townsendii* in northeastern California. *Great Basin Naturalist* 44:431–440.

Percy, D. M., R. D. M. Page, and Q. C. B. Cronk. 2004. Plant-insect interactions: double-dating associated inset and plant lineages reveals asynchronous radiations. *Systematic Biology* 53:120–127.

Pérez, T. M., and W. T. Atyeo. 1984. Feather mites, feather lice, and thanatochresis. *Journal of Parasitology* 70:807–812.

Perotti, M. A., J. M. Allen, D. L. Reed, and H. R. Braig. 2007. Host-symbiont interactions of the primary endosymbiont of human head and body lice. *FASEB Journal* 21:1058–1066.

Perrin, N. 2009. Dispersal. In *The Princeton Guide to Ecology*, edited by S. A. Levin, S. R. Carpenter, H. C. J. Godfray, et al., 45–50. Princeton University Press, Princeton.

Persson, L. 1985. Asymmetrical competition: are larger animals competitively superior? *American Naturalist* 126:261–266.

Peterson, A. T. 1993. Adaptive geographical variation in bill shape of scrub jays (*Aphelocoma coerulescens*). *American Naturalist* 142:508–527.

Pfeffer, A., C. A. Morris, R. S. Green, et al. 2007. Heritability of resistance to infestation with the body louse, *Bovicola ovis*, in Romney sheep bred for differences in resistance or resilience to gastro-intestinal nematode parasites. *International Journal for Parasitology* 37:1589–1597.

Pfennig, D. W., and K. S. Pfennig. 2012. *Evolution's Wedge*. Vol. 12. University of California Press.

Piculell B., J. D. Hoeksema, and J. N. Thompson. 2008. Interactions of biotic and abitoic environmental factors on an ectomycorrhizal symbiosis, and the potential for selection mosaics. *BMC Biology* 6:23.

Pittendrigh, B. R., J. M. Clark, S. H. Lee, W. Sun, and E. Kirkness. 2009. The body louse, *Pediculus humanus humanus* (Phthiraptera: Pediculidae), genome project: past, present, and opportunities for the future. *Urbana* 51:61801.

Poiani, A., A. R. Goldsmith, and M. R. Evans. 2000. Ectoparasites of house sparrows (*Passer domesticus*): an experimental test of the immunocompetence handicap hypothesis and a new model. *Behavioral Ecology and Sociobiology* 47:230–242.

Poinar, G. O., Jr. 2009. Description of an early Cretaceous termite (Isoptera: Kalotermitidae) and its associated intestinal protozoa, with comments on their co-evolution. *Parasites and Vectors* 2:12.

Pollack, R. J., A. Kiszewski, P. Armstrong, et al. Differential permethrin susceptibility of head lice sampled in the United States and Borneo. *Archives of Pediatrics and Adolescent Medicine* 153:969–973.

Pomeroy, D. E. 1962. Birds with abnormal bills. *British Birds* 55:49–72.

Pons, J., A. Hassanin, and P. Crochet. 2005. Phylogenetic relationships within the Laridae (Charadriiformes: Aves) inferred from mitochondrial markers. *Molecular Phylogenetics and Evolution* 37:686–699.

Pope, A. 1713. *Windsor Forest.* Reprinted in: *The Works of Alexander Pope. Poetry. Vol. I.* 1871. John Murray, London.

Poulin, R. 2001. Interactions between species and the structure of helminth communities. *Parasitology* 122:3–11.

———2007. *Evolutionary Ecology of Parasites*. 2nd ed. Princeton University Press, Princeton.

Poulin, R., and M. R. Forbes. 2012. Meta-analysis and research on host-parasite interactions: past and future. *Evolutionary Ecology* 26:1169–1185.

Poulin, R., and S. Morand. 2004. *Parasite Biodiversity*. Smithsonian Books, Washington, DC.

Poulsen B. O. 1994. Poison in Pitohui birds: against predators or ectoparasites? *Emu* 94:128–129.

Prelezov, P., D. Gundasheva, and N. Groseva. 2002. Haematological changes in chickens experimentally infected with biting lice (Phthiraptera-Insecta). *Bulgarian Journal of Veterinary Medicine* 5:29–38.

Prelezov, P. N., N. I. Groseva, and D. I. Goundasheva. 2006. Pathomorphological changes in the tissues of chickens, experimentally infected with biting lice (Insecta:Phthiraptera). *Veterinarski Arhiv* 76:207–215.

Previte, D., B. P. Olds, K. Yoon, et al. 2014. Differential gene expression in laboratory strains of human head and body lice when challenged with *Bartonella quintana*, a pathogenic bacterium. *Insect Molecular Biology* 23:244–254.

Price, M. A., and O. II. Graham. 1997. *Chewing and Sucking Lice as Parasites of Mammals and Birds*. US Department of Agriculture, Technical Bulletin No. 1849, 309 pp.

Price, P. W. 1980. *Evolutionary Biology of Parasites*. Princeton University Press, Princeton.

Price, P. W., M. Westoby, B. Rice, et al. 1986. Parasite mediation in ecological interactions. *Annual Review of Ecology and Systematics* 17:487–505.

Price, R. D., and D. H. Clayton. 1989. *Kaysius emersoni* (Mallophaga: Menoponidae), a new genus and new species of louse from the wedge-billed woodcreeper (Passeriformes: Dendrocolaptidae) of Peru. *Annals of the Entomology Society of America* 82:29–31.

Price, R. D., D. H. Clayton, and R. A. Hellenthal. 1999. Taxonomic review of *Physconelloides* (Phthiraptera: Philopteridae) from the Columbiformes (Aves), with descriptions of three new species. *Journal of Medical Entomology* 36:195–206.

Price, R. D., R. A. Hellenthal, R. L. Palma, K. P. Johnson, and D. H. Clayton. 2003. The chewing lice: world checklist and biological overview. *Illinois Natural History Survey Special Publication 24*.

Price, R. D., and K. P. Johnson. 2009. Five new species of *Myrsidea* Waterston (Phthiraptera:Menoponidae) from tanagers (Passeriformes:Thraupidae) in Panama. *Zootaxa* 2200:61–68.

Priestley, C. M., I. F. Burgess, and E. M. Williamson. 2006. Lethality of essential oil constituents towards the human louse, *Pediculus humanus*, and its eggs. *Fitoterapia* 77:303–309.

Proctor, N. S., and P. J. Lynch. 1993. *Manual of Ornithology: Avian Structure and Function*. Yale University Press, New Haven.

Råberg, L., D. Sim, and A. F. Read. 2007. Disentangling genetic variation for resistance and tolerance to infectious diseases in animals. *Science* 318:812–814.

Råberg, L., J. De Roode, A. Bell, P. Stamou, D. Gray, and A. Read. 2006. The role of immune-mediated apparent competition in genetically diverse malaria infections. *American Naturalist* 168:41–53.

Radford, A. N., and M. A. Du Plessis. 2006. Dual function of allopreening in the cooperatively breeding green woodhoopoe, *Poeniculus purpueus*. *Behavioral Ecology and Sociobiology* 61:221–230.

Raghavan, R. S., K. R. Reddy, and G. A. Khan. 1968. Dermatitis in elephants caused by the louse *Haematomyzus elephantis* (Piagot [*sic*] 1869). *Indian Veterinary Journal* 45: 700–701.

Rand, D. M. 1994. Thermal habit, metabolic rate and the evolution of mitochondrial DNA. *Trends in Ecology and Evolution* 9:125–131.

Rannala, B., and Y. Michalakis. 2003. Population genetics and cospeciation: from process to pattern. In *Tangled Trees: Phylogeny, Cospeciation and Coevolution*, edited by R. D. M. Page, 120–143. University of Chicago Press, Chicago.

Rantala, M. J. 1999. Human nakedness: adaptation against ectoparasites? *International Journal for Parasitology* 29:1987–1989.

Raoult, D., O. Dutour, L. Houhamdi, et al. 2006. Evidence for louse-transmitted diseases in soldiers of Napoleon's Grand Army in Vilnius. *Journal of Infectious Diseases* 193:112–120.

Rattenborg, N. C. 2006. Do birds sleep in flight? *Naturwissenschaften* 93:413–425.

Ratzlaff, R. E., and S. K. Wikel. 1990. Murine immune responses and immunization against *Polyplax serrata* (Anoplura: Polyplacidae). *Journal of Medical Entomology* 27:1002–1007.

Redpath, S. 1988. Vigilance levels in preening Dunlin *Caladris alpina*. *Ibis* 130:555–557.

Reed, D. L., R. W. Currier, S. F. Walton, et. al. 2011. The evolution of infectious agents in relation to sex in animals and humans: brief discussions of some individual organisms. *Annals of the New York Academy of Sciences* 1230:74–107.

Reed, D. L., and M. S. Hafner. 1997. Host specificity of chewing lice on pocket gophers: a potential mechanism for cospeciation. *Journal of Mammalogy* 78:655–660.

Reed, D. L., M. S. Hafner, and S. K. Allen. 2000a. Mammal hair diameter as a possible mechanism for host specialization in chewing lice. *Journal of Mammalogy* 81: 999–1007.

Reed, D. L., M. S. Hafner, S. K. Allen, and M. B. Smith. 2000b. Spatial partitioning of host habitat by chewing lice of the genera *Geomydoecus* and *Thomomydoecus* (Phthiraptera: Trichodectidae). *Journal of Parasitology* 86:951–955.

Reed, D. L., J. E. Light, J. M. Allen, and J. J. Kirchman. 2007. Pair of lice lost or parasites regained: the evolutionary history of anthropoid primate lice. *BMC Biology* 5:7.

Reed D. L., V. S. Smith, A. R. Rogers, S. L. Hammond, and D. H. Clayton. 2004. Molecular genetic analysis of human lice supports direct contact between modern and archaic humans. *PLoS Biology* 2:1972–1983.

Reed, D. L., M. A. Toups, J. A. Light, J. M. Allen, and S. Flannigan. 2009. Lice and other parasites as markers of primate evolutionary history. In *Primate Parasite Ecology: The Dynamics and Study of Host-Parasite Relationships*, edited by M. A. Huffman and C. A. Chapman, 231–250.Cambridge University Press, Cambridge.

Reeves, W. K., M. P. Nelder, and J. A. Korecki. 2005. *Bartonella* and *Rickettsia* in fleas and lice from mammals in South Carolina, USA. *Journal of Vector Ecology* 30: 310–315.

Rékási, J., L. Rózsa, and B. J. Kiss. 1997. Patterns in the distribution of avian lice (Phthiraptera:Amblycera, Ischnocera). *Journal of Avian Biology* 28:150–156.

Reznick, D. N., and R. E. Ricklefs. 2009. Darwin's bridge between microevolution and macroevolution. *Nature* 457:837–842.

Ribeiro, J. M. C, and I. M. B. Francischetti. 2003. Role of arthropod saliva in blood feeding: sialome and post-sialome perspectives. *Annual Review of Entomology* 48: 73–88.

Ribeiro, J. M. C., G. T. Makoul, J. Levine, D. R. Robinson, and A. Spielman. 1985. Antihemostatic, antiinflammatory, and immunosuppressive properties of the saliva of a tick, *Ixodes dammini*. *Journal Experimental Medicine* 161:332–344.

Ricklefs, R. E. 2010a. Evolutionary diversification, coevolution between populations and their antagonists, and the filling of niche space. *Proceedings of the National Academy of Sciences USA* 107:1265–1272.

———. 2010b. Host-pathogen coevolution, secondary sympatry and species diversification. *Philosophical Transactions of the Royal Society* 365:1139–1147.

Ricklefs, R. E, and E. Bermingham. 2008. The West Indies as a laboratory of biogeography and evolution. *Philosophical Transactions of the Royal Society B* 363: 2393–2413.

Ries, E. 1931. Die Symbiose der Lause und Federlinge. *Zeitschrift für Morphologie und Ökologie der Tiere* 20:233–367.

Ritter, R. C., and A. N. Epstein. 1974. Saliva lost by grooming: a major item in the rat's water economy. *Behavioral Biology* 11:581–585.

Roberts, L. S. 2000. The crowding effect revisited. *Journal of Parasitology* 86:209–211.

Roberts, R. J. 2002. Head lice. *New England Journal of Medicine* 346:1645–1650.

Rodgman, A., and T. A. Perfetti. 2008. *The Chemical Components of Tobacco and Tobacco Smoke*. CRC, Boca Raton, FL.

Rodríguez-Gironés, M. A., and L. Santamaría. 2007. Resource competition, character displacement, and the evolution of deep corolla tubes. *American Naturalist* 170: 455–464.

Rohde, K. 1979. A critical evaluation of intrinsic and extrinsic factors responsible for niche restriction in parasites. *American Naturalist* 114:648–671.

Ronald, N. C., and J. E. Wagner. 1973. Pediculosis of spider monkeys: a case report with zoonotic implications. *Laboratory Animal Science* 23:872.

Ronquist, F. 2003. Parsimony analysis of coevolving species associations. In *Tangled Trees: Phylogeny, Cospeciation, and Coevolution*, edited by R. D. M. Page, 22–64. University of Chicago Press, Chicago.

Rose, K. D., A. Walker, and L. L. Jacobs. 1981. Function of the mandibular tooth comb in living and extinct mammals. *Nature* 289:583–585.

Rossini, C., L. Castillo, and A. González. 2008. Plant extracts and their components as potential control agents against human head lice. *Phytochemistry Reviews* 7: 51–63.

Rothschild, M., and T. Clay. 1957. *Fleas, Flukes and Cuckoos*. Macmillan, New York.

Rothschild, M., and R. Ford. 1964. Breeding of the rabbit flea (*Spilopsyllus cuniculi* [Dale]), controlled by the reproductive hormones of the host. *Nature* 201:103–104.

Rothschild, W., and K. Jordan. 1903. *A Revision of the Lepidopterous Family Sphingidae*. Hazell, Watson and Viney, London.

Rózsa, L. 1991. Points in question: flamingo lice contravene Fahrenholz. *International Journal for Parasitology* 21:151–152.

———. 1997. Patterns in the abundance of avian lice (Phthiraptera:Amblycera, Ischnocera). *Journal of Avian Biology* 28:249–254.

Rózsa, L., and P. Apari. 2012. Why infest the loved ones—inherent human behaviour indicates former mutualism with head lice. *Parasitology* 139:696–700.

Rózsa, L., J. Rékási, and J. Reiczigel. 1996. Relationship of host coloniality to the population ecology of avian lice (Insecta: Phthiraptera). *Journal of Animal Ecology* 65:242–248.

Rudolph, D. 1983. The water-vapour uptake system of the Phthiraptera. *Journal of Insect Physiology* 29:15–25.

Rueesch, S., M. Lemoine, and H. Richner. 2012. Ectoparasite reproductive performance when host condition varies. *Parasitology Research* 111:1193–1203. doi: 10.1007/s00436-012-2953-3.

Rundell, R. J., and T. D. Price. 2009. Adaptive radiation, nonadaptive radiation, ecological speciation and nonecological speciation. *Trends in Ecology and Evolution* 24:394–399.

Rundle, P. A., and D. S. Hughes. 1993. *Phthirus pubis* infestation of the eyelids. *British Journal of Ophthalmology* 77:815–816.

Rust, R. W. 1974. The population dynamics and host utilization of *Geomydoecus oregonus*, a parasite of *Thomomys bottae*. *Oecologia* 3:287–304.

Ruxton, G. D., T. N. Sherratt, and M. P. Speed. 2004. *Avoiding Attack: The Evolutionary Ecology of Crypsis, Warning Signals and Mimicry*. Oxford University Press, Oxford.

Sadd, B. M. 2011. Food-environment mediates the outcome of specific interactions between a bumblebee and its trypanosome parasite. *Evolution* 65:2995–3001.

Sage, R. D., D. Heyneman, K.-C. Lim, and A. C. Wilson. 1986. Wormy mice in a hybrid zone. *Nature* 324:60–63.

Saino, N., and A. P. Møller. 1994. Secondary sexual characters, parasites and testosterone in the barn swallow, *Hirundo rustica*. *Animal Behavior* 48:1325–1333.

Saino, N., A. P. Møller, and A. M. Bolzerna. 1995. Testosterone effects on the immune system and parasite infestations in the barn swallow (*Hirundo rustica*): an experimental test of the immunocompetence hypothesis. *Behavioral Ecology* 6: 397–404.

Samuel, W. M., E. S. Williams, and A. B. Rippin. 1982. Infestations of *Piagetiella peralis* (Mallophaga: Menoponidae) on juvenile white pelicans. *Canadian Journal of Zoology* 60:951–953.

Sandre, S.-L., A. Kaasik, U. Eulitz, and T. Tammaru. 2013. Phenotypic plasticity in a generalist insect herbivore with the combined use of direct and indirect cues. *Oikos* 122:1626–1635.

Sangster, N. C. 2001. Managing parasiticide resistance. *Veterinary Parasitology* 98: 89–109.

Sari, E. H. R., H. Klompen, and P. G. Parker. 2013. Tracking the origins of lice, haemosporidian parasites and feather mites of the Galapagos flycatcher (*Myiarchus magnirostris*). *Journal of Biogeography* 40:1082–1093.

Sasal, P., N. Niquil, and P. Bartoli. 1999. Community structure of digenean parasites of sparid and labrid fishes of the Mediterranean sea: a new approach. *Parasitology* 119:635–648.

Sasaki-Fukatsu, K., R. Koga, N. O. Nikoh, et al. 2006. Symbiotic bacteria associated with stomach discs of human lice. *Applied and Environmental Microbiology* 72: 7349–7352.

Saxena, A. K., G. P. Agarwal, S. Chandra, and O. P. Singh. 1985. Pathogenic involvement of Mallophaga. *Zeitschrift für Angewandte Entomologie* 99:294–301.

Sazima, I. 2011. Cleaner birds: a worldwide overview. *Revista Brasileira de Ornitologia* 19:32–47.

Scheiner, S. M. 2014. The Baldwin effect: neglected and misunderstood. *American Naturalist* 184:ii–iii.

Schluter, D. 2000a. Ecological character displacement in adaptive radiation. *American Naturalist* 156:S4–16.

———. 2000b. *The Ecology of Adaptive Radiation*. Oxford University Press, Oxford.

———. 2009. Evidence for ecological speciation and its alternative. *Science* 323: 737–741.

———. 2010. Resource competition and coevolution in sticklebacks. *Evolution: Education and Outreach* 3:54–61.

Schmid-Hempel, P. 2011. *Evolutionary Parasitology: The Integrated Study of Infections, Immunology, Ecology, and Genetics*. Oxford University Press, New York.

Schmidt, A. R., S. Jancke, E. E. Lindquist, et al. 2012. Arthropods in amber from the Triassic Period. *Proceedings of the National Academy of Sciences USA* 109:14796–14801.

Schulte, R. D., C. Makus, and H. Schulenburg. 2013. Host–parasite coevolution favours parasite genetic diversity and horizontal gene transfer. *Journal of Evolutionary Biology* 26:1836–1840.

Schwartz, C. C., R. Stephenson, and N. Wilson. 1983. *Trichodectes canis* on the gray wolf and coyote on Kenai Peninsula, Alaska. *Journal of Wildlife Diseases* 19:372–373.

Seegar, W. S., E. L. Schiller, W. J. L. Sladen, and M. Trpis. 1976. A Mallophaga, *Trinton anserium*, as a cyclodevelopmental vector for a heartworm parasite of waterfowl. *Science* 194:739–741.

Seehausen, O. 2004. Hybridization and adaptive radiation. *Trends in Ecology and Evolution* 19:198–207.

Seger, J., W. A. Smith, J. J. Perry, et al. 2010. Gene genealogies strongly distorted by weakly interfering mutations in constant environments. *Genetics* 184:529–545.

Segraves, K. A. 2010. Branching out with coevolutionary trees. *Evolution: Education and Outreach* 3:62–70.

Servedio, M. R., G. S. Van Doorn, M. Kopp, A. M. Frame, and P. Nosil. 2011. Magic traits in speciation: "magic" but not rare? *Trends in Ecology and Evolution* 26: 389–397.

Shapiro, M. D., and E. T. Domyan. 2013. Domestic pigeons. *Current Biology* 23:R302.

Sheldon, B. C., and S. Verhulst. 1996. Ecological immunology:costly parasite defenses and trade offs in evolutionary ecology. *Trends in Ecology and Evolution* 11: 317–321.

Sibly, R. M., and P. Calow. 1986. Physiological ecology of animals. Blackwell Scientific, Oxford.

Simková, A., Y. Desdevises, M. Gelnar, and S. Morand. 2000. Co-existence of nine gill ectoparasites (Dactylogyrus: monogenea) parasitising the roach (*Rutilus rutilus* l.): history and present ecology. *International Journal for Parasitology* 30:1077–1088.

Simková, A., S. Morand, E. Jobet, M. Gelnar, and O. Verneau. 2004. Molecular phylogeny of congeneric monogenean parasites (*Dactylogryus*): a case of intrahost speciation. *Evolution* 58:1001–1018.

Simmons, K. E. L. 1966. Anting and the problem of self-stimulation. *Journal of Zoology* 149:145–162.

———. 1986. *The Sunning Behaviour of Birds*. Bristol Ornithological Club, Bristol.

Simmons, R. B., and S. J. Weller. 2001. Utility and evolution of cytochrome b in insects. *Molecular Phylogenetics and Evolution* 20:196–210.

Simpson, G. G. 1944. *Tempo and Mode in Evolution*. Columbia Biological Series No. 15. Columbia University Press, New York.

———. 1953. *The Major Features of Evolution*. Columbia University Press, New York.

Singer, M. S., and J. O. Stireman. 2005. The tri-trophic niche concept and adaptive radiation of phytophagous insects. *Ecology Letters* 8:1247–1255.

Skirnisson, K., and E. Olafsson. 1990. Parasites of seals in Icelandic waters, with special reference to the heartworm *Dipetalonema spirocauda* Leidy, 1858 and the sucking louse *Echinopthirius* [*sic*] *horridus* Olfers, 1916. *Natturufraedingurinn* 60: 93–102.

Smith, C. I., W. K. Godsoe, S. Tank, J. B. Yoder, and O. Pellmyr. 2008. Distinguishing coevolution from covicariance in an obligate pollination mutualism: asynchronous divergence in Joshua tree and its pollinators. *Evolution* 62:2676–2687.

Smith, J. W., and C. W. Benkman. 2007. A coevolutionary arms race causes ecological speciation in crossbills. *American Naturalist* 169:455–465.

Smith, V. S. 2001. Avian louse phylogeny (Phthiraptera: Ischnocera): a cladistic study based on morphology. *Zoological Journal of the Linnean Society* 132:81–144.

Smith, V. S., T. Ford, K. P. Johnson, P. C. Johnson, K. Yoshizawa, and J. E. Light. 2011. Multiple lineages of lice pass through the K–Pg boundary. *Biology Letters* 7: 782–785.

Smith, V. S., J. E. Light, and L. A. Durden. 2008. Rodent louse diversity, phylogeny, and cospeciaiton in the Manu Biosphere Reserve, Peru. *Biological Journal of the Linnean Society* 95:598–610.

Smith, W. A. 2011. Bacterial symbionts of the feather louse genus *Columbicola*: characterization and phylogenetic analyses. PhD diss., University of Utah.

Smith, W. A., C. Dale, and D. H. Clayton. 2010. Determining the role of bacterial symbionts with the genus *Columbicola*. *Turkiye Parazitologia Derg* 34:67.

Smith, W. A., K. P. Johnson, D. L. Reed, et al. 2013. Phylogenetic analysis of symbionts in feather-feeding lice of the genus *Columbicola*: evidence for repeated symbiont replacements. *BMC Evolutionary Biology* 13:109–123.

Sobel, J. M., G. F. Chen, L. R. Watt, and D. W Schemske. 2010. The biology of speciation. *Evolution* 64:295–315.

Soler, J. J., J. M. Peralta-Sánchez, A. M. Martín-Platero, M. Martín-Vivaldi, M. Martínez-Bueno, and A. P. Møller. 2012. The evolution of size of the uropygial gland: mutualistic feather mites and uropygial secretion reduce bacterial loads of eggshells and hatching failures of European birds. *Journal of Evolutionary Biology* 25:1779–1791.

Sorenson, M. D., C. N. Balakrishnan, and R. B. Payne. 2004. Clade-limited colonization in brood parasitic finches (*Vidua* spp.). *Systematic Biology* 53:140–153.

Spurrier, M. F., M. S. Boyce, and B. F. J. Manly. 1991. *Bird-Parasite Interactions: Ecology, Evolution and Behaviour*, edited by J. E. Loye and M. Zuk, 389–398. Oxford University Press, Oxford.

Stammer, H. J. 1957. Gedanken zu den parasitophyletischen Regeln und zur Evolution der Parasiten. *Zoologischer Anzeiger* 159:255–267.

Štefka, J., P. E. A. Hoeck, L. F. Keller, and V. S. Smith. 2011. A hitchhikers guide to the Galápagos: co-phylogeography of Galápagos mockingbirds and their parasites. *BMC Evolutionary Biology* 11:284.

Stenram, H. 1956. The ecology of *Columbicola columbae* L. (Mallophaga). *Opuscula Entomologica* 21:170–190.

Stevens, M., and S. Merilaita. 2009. Animal camouflage: current issues and new perspectives. *Philosophical Transactions of the Royal Society B* 364:423.

Stewart, S. J., J. F. Bell, B. Hestekin, and G. J. Moore. 1976. Effects of limb disability on lousiness in mice. VI. Lack of tolerance after neonatal exposure. *Experimental Parasitology* 40:373–379.

Stiling, P., D. Simberloff, and B. V. Brodbeck. 1991. Variation in rates of leaf abscission between plants may affect the distribution patterns of sessile insects. *Oecologia* 88:367–370.

Stock, T. M., and J. C. Holmes. 1988. Functional relationships and microhabitat distributions of enteric helminths of grebes (Podicipedidae): the evidence for interactive communities. *Journal of Parasitology* 74:214–227.

Stockdale, H. J., and E. S. Raun. 1960. Economic importance of the chicken body louse. *Journal of Economic Entomology* 53:421–423.

Stone, L., and A. Roberts. 1991. Conditions for a species to gain advantage from the presence of competitors. *Ecology* 72: 1964–1972.

Stone, W. B. 1969. The ecology of parasitism in captive waterfowl at the Burnet Park Zoo. PhD diss., Syracuse University.

Strong, D. R., J. H. Lawton, and S. R. Southwood. 1984. *Insects on Plants: Community Patterns and Mechanisms.* Blackwell Scientific Publicatons, Oxford.

Suárez-Rodríguez, M., I. López-Rull, and C. M. Garcia. 2013. Incorporation of cigarette butts into nests reduces nest ectoparasite load in urban birds: new ingredients for an old recipe? *Biology Letters* 9. doi:10.1098/rsbl.2012.0931.

Subbotin, S. A., E. L. Krall, I. T. Riley, et al. 2004. Evolution of the gall-forming plant parasitic nematodes (Tylenchida: Anguinidae) and their relationships with hosts as inferred from Internal Transcribed Spacer sequences of nuclear ribosomal DNA. *Molecular Phylogenetics and Evolution* 30:226–235.

Suwa, G., R. T. Kono, S. Katoh, B. Asfaw, and Y. Beyene. 2007. A new species of great ape from the late Miocene epoch in Ethiopa. *Nature* 448:921–924.

Svensson, E. I., and L. Råberg. 2010. Resistance and tolerance in animal enemy-victim coevolution. *Trends in Ecology and Evolution* 25:267–274.

Switzer, W. M., M. Salemi, V. Shanmugam, et al. 2005. Ancient co-speciation of simian foamy viruses and primates. *Nature* 434:376–380.

Takiya, D. M., P. L. Tran, C. H. Dietrich, and N. A. Moran. 2006. Co-cladogenesis spanning three phyla: leafhoppers (Insecta: Hemiptera: Cicadellidae) and their dual bacterial symbionts. *Molecular Ecology* 15:4175–4191.

Talbert, R., and R. Wall. 2012. Toxicity of essential and non-essential oils against the chewing louse, *Bovicola* (*Werneckiella*) *ocellatus*. *Research in Veterinary Science* 93: 831–835.

Tanaka, I. 1995. Matrilineal distribution of louse egg-handling techniques during grooming in free-ranging Japanese macaques. *American Journal of Physical Anthropology* 98:197–201.

———. 1998. Social diffusion of modified louse egg-handling techniques during grooming in free-ranging Japanese macaques. *Animal Behavior* 56:1229–1236.

Tanaka, I., and H. Takefushi. 1993. Elimination of external parasites (lice) is the primary function of grooming in free-ranging Japanese Macaques. *Anthropological Science* 101:187–193.

Taper, M. L., and T. J. Case. 1992. Coevolution among competitors. *Oxford Surveys in Evolutionary Biology* 8:63–109.

Telfer, S., X. Lambin, R. Birtles, et al. 2010. Species interactions in a parasite community drive infection risk in a wildlife population. *Science* 330:243–246.

Tellier, A., D. M. de Vienne, T. Giraud, M. E. Hood, and G. Refrégier. 2010. Theory and examples of reciprocal influence between hosts and pathogens, from short-term to long-term interactions: coevolution, cospeciation and pathogen speciation. In *Host-Pathogen Interactions: Genetics, Immunology and Physiology*, edited by A. W. Barton, 37–77. Nova Science, Hauppauge, NY.

Tello, J. S., R. D. Stevens, and C. W. Dick. 2008. Patterns of species co-occurrence and density compensation: a test for interspecific competition in bat ectoparasite infracommunities. *Oikos* 117:693–702.

Thompson, G. B. 1940. The distribution of *Heterodoxus spiniger* (Enderlein). *Papers and Proceedings of the Royal Society of Tasmania* 1939:27–31.

Thompson, J. N. 1982. *Interaction and Coevolution*. Wiley, New York.

———. 1987. Symbiont-induced speciation. *Biological Journal of the Linnean Soeciety* 32:385–393.

———. 1989. Concepts of coevolution. *Trends in Ecology and Evolution* 4:179–183.

———. 1994. *The Coevolutionary Process*. University of Chicago Press, Chicago.

———. 2005. *The Geographic Mosaic of Coevolution*. University of Chicago Press, Chicago.

———. 2009. The coevolving web of life (American Society of Naturalists Presidential Address). *American Naturalist* 173:125–140.

———. 2013. *Relentless Evolution.* University of Chicago Press, Chicago.

———. 2014. Natural selection, coevolution and the web of life. *American Naturalist* 183.

Thompson, J. N., and B. M. Cunningham. 2002. Geographic structure and dynamics of coevolutionary selection. *Nature* 417:735–738.

Thomson, J. D. 1980. Implications of different sorts of evidence for competition. *American Naturalist* 116:719–726.

Tilman, D. 1994. Competition and biodiversity in spatially structured habitats. *Ecology* 75:2–16.

Timm, R. M. 1983. Fahrenholz's rule and resource tracking: a study of host-parasite coevolution. In *Coevolution*, edited by M. H. Nitecki, 225–265. University of Chicago Press, Chicago.

Timmermann, G. 1952. The species of the genus *Quadraceps* (Mallophaga) from the Larinae. With some remarks on the systematic and the phylogeny of the gulls. *Annals and Magazine of Natural History* 51:209–223.

Toju, H. 2009. Natural selection drives the fine-scale divergence of a coevolutionary arms race involving a long-mouthed weevil and its obligate host plant. *BMC Evolutionary Biology* 9:273.

Toju, H. 2011. Weevils and camellias in a Darwin's race: model system for the study of eco-evolutionary interactions between species. *Ecological Research* 26:239–251.

Toju, H., and T. Sota. 2006. Phylogeography and the geographic cline in the armament of a seed-predatory weevil: effects of historical events vs. natural selection from the host plant. *Molecular Ecology* 15:4161–4173.

Tompkins, D. M., and D. H. Clayton. 1999. Host resources govern the specificity of swiftlet lice: size matters. *Journal of Animal Ecology* 68:489–500.

Tompkins, D. M., T. Jones, and D. H. Clayton. 1996. Effect of vertically transmitted ectoparasites on the reproductive success of swifts (*Apus apus*). *Functional Ecology* 10:733–740.

Toon, A., and J. M. Hughes. 2008. Are lice good proxies for host history? A comparative analysis of the Australian magpie, *Gymnorhina tibicen*, and two species of feather louse. *Heredity* 101:127–135.

Torchin, M. E., K. D. Lafferty, A. P. Dobson, V. J. McKenzie, and A. M. Kuris. 2003. Introduced species and their missing parasites. *Nature* 421:628–630.

Toups, M. A., A. Kitchen, J. E. Light, and D. L. Reed. 2011. Origin of clothing lice indicates early clothing use by anatomically modern humans in Africa. *Molecular Biology and Evolution* 28:29–32.

Trontelj, P., A. Blejec, and C. Fiser. 2012. Ecomorphological convergence of cave communities. *Evolution* 66:3852–3865.

True, T. E. 2003. Insect melanism: the molecules matter. *Trends in Ecology and Evolution* 18:640–647.

Tschirren, B., L. L. Bischoff, V. Saladin, and H. Richner. 2006. Host condition and host immunity affect parasite fitness in a bird-ectoparasite system. *Functional Ecology* 21:372–378.

Tschirren, B., P. S. Fitze, and H. Richner. 2007. Maternal modulation of natal dispersal in a passerine bird: an adaptive strategy to cope with parasitism? *American Naturalist* 169:87–93.

Turner, M. L. 2003. The micromorphology of the blesbuck louse *Damalinia crenelata* as observed under the scanning electron microscope. *Koedoe-African Protected Area Conservation and Science* 46:65–71.

Turner, M. L., C. Labuschagne, and E. D. Green. 2004. The micromorphology of the African buffalo louse *Haematopinus bufali* as observed under the scanning electron microscope. *Koedoe-African Protected Area Conservation and Science* 47: 83–90.

Vágási, C. I. 2014. The origin of feather holes: a word of caution. *Journal of Avian Biology* (in press).

van der Cingel, N. A. 2001. *An Atlas of Orchid Pollination*. A. A. Balkema, Rotterdam.

Van Liere, D.W. 1992. The significance of fowl's bathing in dust. *Animal Welfare* 1: 187–202.

Van Valen, L. 1983. How pervasive is coevolution? In *Coevolution*, edited by M. H. Nitecki, 1–19. University of Chicago Press, Chicago.

Van Veen, F. J. F., R. J. Morris, and H. C. J. Godfray. 2006. Apparent competition, quantitative food webs, and the structure of phytophagous insect communities. *Annual Review of Entomology* 51:187–208.

Vas, Z., G. Csorba, and L. Rózsa. 2012. Evolutionary co-variation of host and parasite diversity—the first test of Eichler's rule using parasitic lice (Insecta: Phthiraptera). *Parasitology Research* 111:393–401.

Vas Z, T. Csörg, A. P. Møller, and L. Rozsa. 2008. The feather holes on the barn swallow *Hirundo rustica* and other small passerines are probably caused by *Brueelia* spp. lice. *Journal of Parasitology* 94:1438–1440.

Vas, Z., L. Lefebvre, K. P. Johnson, J. Reiczigel, and L. Rózsa. 2011. Clever birds are lousy: co-variation between avian innovation and the taxonomic richness of their amblyceran lice. *International Journal for Parasitology* 41:1295–1300.

Veracx, A., A. Boutellis, V. Merhej, G. Diatta, and D. Raoult. 2012. Evidence for an African cluster of human head and body lice with variable colors and interbreeding of lice between continents. *PLoS ONE* 7:e37804.

Veracx, A., and D. Raoult. 2012. Biology and genetics of human head and body lice. *Trends in Parasitology* 28:563–571.

Verderane, M. P., T. Falótico, B. D. Resende, M. B. Labruna, P. Izar, and E. B. Ottoni. 2007. Anting in a semifree-ranging group of *Cebus apella*. *International Journal of Primatology* 28:47–53.

Vermeij, G. J. 1983. Intimate associations and coevolution in the seas. In *Coevolution*, edited by D. J. Futuyma and M. Slatkin, 311–327. Sinauer, Sunderland, MA.

Vermeij, G. J. 2008. Escalation and its role in Jurrasic biotic history. *Palaeogeography, Palaeoclimatology, Palaeoecology* 263:3–8.

Viblanc, V. A., A. Mathien, C. Saraux, V. M. Viera, and R. Groscolas. 2011. It costs to be clean and fit: energetics of comfort behavior in breeding-fasting penguins. *PloS ONE* 6:e21110.

Vidal-Martínez, V. M., and C. R. Kennedy. 2000. Potential interactions between the intestinal helminths of the cichlid fish *Cichlasoma synspilum* from southeastern Mexico. *Journal of Parasitology* 86:691–695.

Vogwill, T., A. Fenton, and M. A. Brockhurst. 2010. How does spatial dispersal network affect the evolution of parasite local adaptation? *Evolution* 64:1795–1801.

Volf, P. 1991. *Polyplax spinulosa* infestation and antibody response in various strains of laboratory rats. *Folia Parasitologica* 38:355.

———. 1994. Localization of the major immunogen and other glycoproteins of the louse *Polyplax spinulosa*. *International Journal for Parasitology* 24:1005–1010.

Waage, J. K. 1979. The evolution of insect/vertebrate associations. *Biological Journal of the Linnean Society* 12:187–224.

Waage, J. K., and C. R. Davies. 1986. Host-mediated competition in a bloodsucking insect community. *Journal of Animal Ecology* 55:171–180.

Wade, M. J. 2007. The coevolutionary genetics of ecological communities. *Nature Reviews Genetics* 8:185–195.

Waite, J. L., A. R. Henry, and D. H. Clayton. 2012. How effective is preening for controlling mobile ectoparasites? An experimental test with pigeons and hippoboscid flies. *International Journal of Parasitology* 42:463–467.

Wakelin, D., and V. Apanius. 1997. Immune defense:genetic control. In *Host-Parasite Evolution: General Principles and Avian Models*, edited by D. H. Clayton and J. Moore, 30–58.Oxford University Press, Oxford.

Wallace, A. R. 1893. Preface. In *The dispersal of shells: An inquiry into the means of dispersal possessed by fresh-water and land Mollusca*, Vol. 75, edited by H. W. Kew. Kegan Paul, Trench, Trübner.

Walther, B., and D. Clayton. 2005. Elaborate ornaments are costly to maintain: evidence for high maintenance handicaps. *Behavioral Ecology* 16:89.

Wappler, T., V. S. Smith, and R. C. Dalgleish. 2004. Scratching an ancient itch: an Eocene bird louse fossil. *Proceedings of the Royal Society of London B* 271: S255–258.

Ward, R. A. 1957. A study of the host distribution and some relationships of biting lice (Mallophaga) parasitic on birds of the order Tinamiformes. II. *Annals of the Entomological Society of America* 50:452–459.

Waterhouse, D. F. 1957. Digestion in insects. *Annual Review of Entomology* 2:1–18.

Watson, D. W. D., J. E. J. Lloyd, and R. R. Kumar. 1997. Density and distribution of cattle lice (Phthiraptera: Haematopinidae, Linognathidae, Trichodectidae) on six steers. *Veterinary Parasitology* 69:283–296.

Weber, M. G., and A. A. Agrawal. 2012. Phylogeny, ecology, and the coupling of comparative and experimental approaches. *Trends in Ecology and Evolution* 27: 394–403.

Weckstein, J. D. 2004. Biogeography explains cophylogenetic pattersn in toucan chewing lice. *Systematic Biology* 53:154–164.

Weiblen, G. D., and G. L. Bush. 2002. Speciation in fig pollinators and parasites. *Molecular Ecology* 11:1573–1578.

Weilgama, D. J. 1999. Ectoparasites. In *Water Buffalo in Asia, Vol. 4: Diseases of the Buffalo*, edited by M. C. L. De Alwis, D. H. A. Subasinghe, and N. U. Horadagoda, 27–35. National Science Foundation, Arlington.

Weiss, R. A. 2009. Apes, lice and prehistory. *Journal of Biology* 8:20.

Weldon, P. J., and J. F. Carroll. 2006. Vertebrate chemical defense: secreted and topically acquired deterrents of arthropods. In *Insect Repellents: Principles, Methods and Uses*, edited by M. Debboun, S. P. Frances, and D. Stickman, 47–75. CRC, New York.

Weldon, P. J., J. F. Carroll, M. Kramer, R. H. Bedoukian, R. E. Coleman, and U. R. Bernier. 2011. Anointing chemicals and hematophagous arthropods: responses by ticks and mosquitoes to citrus (Rutaceae) peel exudates and monoterpene components. *Journal of Chemical Ecology* 37:348–359.

Westrom, D., and R. Yescott. 1975. Emigration of ectoparasites from dead California ground squirrels *Spermophilus beecheyi* (Richardson). *California Vector Views* 22: 97–103.

Westrom, D. R., B. C. Nelson, and G. E. Connolly. 1976. Transfer of *Bovicola tibialis* (Piaget) (Mallophaga: Trichodectidae) from the introduced fallow deer to the Columbian black-tailed deer in California. *Journal of Medical Entomology* 13: 169–173.

Wetmore, A. 1936. The number of contour feathers in passeriform and related birds. *Auk* 53:159–169.

Whitaker, L. M. 1957. A résumé of anting, with particular reference to a captive orchard oriole. *Wilson Bulletin* 69:195–262.

Whiteman, N. K. 2008. Lice help illuminate the recent evolutionary history of an Australian bird. *Heredity* 101:105–106.

Whiteman, N. K., V. S. Dosanjh, R. L. Palma, et al. 2009. Molecular and morphological divergence in a pair of bird species and their ectoparasites. *Journal of Parasitology* 95:1372–1382.

Whiteman, N. K., R. T. Kimball, and P. G. Parker. 2007. Co-phylogeography and comparative population genetics of the threatened Galapagos hawk and three ectoparasite species: ecology shapes population histories within parasite communities. *Molecular Ecology* 22:4759–4773.

Whiteman, N. K., K. D. Matson, J. L. Bollmer, and P. G. Parker. 2006. Disease ecology in the Galápagos hawk (*Buteo galapagoensis*): host genetic diversity, parasite load and natural antibodies. *Proceedings of the Royal Society of London B* 273:797–804.

Whiteman, N. K., and P. G. Parker. 2004. Effects of host sociality on ectoparasite population biology. *Journal of Parasitology* 90:939–947.

———. 2005. Using parasites to infer host population history: a new rationale for parasite conservation. *Animal Conservation* 8:175–181.

Whiteman, N. K., D. Santiago-Alarcon, K. P. Johnson, and P. G. Parker. 2004. Differences in straggling rates between two genera of dove lice (Insecta: Phthiraptera) reinforce population genetic and cophylogenetic patterns. *International Journal for Parasitology* 34:1113–1119.

Wible, J. R., G. W. Rougier, M. J. Novacek, and R. J. Asher. 2007. Cretaceous eutherians and Laurasian origin for placental mammals near the K/T boundary. *Nature* 447:1003–1006. doi:10.1038/nature05854.

Wikel, S. K. 1982. Immune responses to arthropods and their products. *Annual Review of Entomology* 27:21–48.

———. 1984. Immunomodulation of host responses to ectoparasite infestation—an overview. *Veterinary Parasitology* 14:321–339.

———. 1996. *The Immunology of Host-Ectoparasitic Arthropod Relationships*. CAB International, Wallingford, UK.

———. 1999. Modulation of the host immune system by ectoparasitic arthropods. *BioScience* 49:311–320.

Wilson, F. H. 1937. Lice on hibernating and non-hibernating mammals. *Journal of Mammalogy* 18:361–362.

Windsor, D. A. 1998. Most of the species on Earth are parasites. *International Journal for Parasitology* 28:1939–1941.

Winkler, I. S., and C. M. Mitter. 2008. The phylogenetic dimension of insect/plant assemblages: a review of recent evidence. In *Specialization, Speciation, and Radiations: The Evolutionary Biology of Herbivorous Insects*, edited by K. J. Tilmon, 240–263. University of California Press, Berkeley.

Winkler, I. S., C. M. Mitter, and S. J. Scheffer. 2009. Repeated climate-linked host shifts have promoted diversification in a temperate clade of leaf-mining flies. *Proceedings of the National Academy of Sciences USA* 106:18103–18108.

Wolff, E. D., S. W. Salisbury, J. R. Horner, and D. J. Varricchio. 2009. Common avian infection plagued the tyrant dinosaurs. *PloS ONE* 4:e7288.

Woolhouse, M. E., J. P. Webster, E. Domingo, B. Charlesworth, and B. R. Levin. 2002. Biological and biomedical implications of the co-evolution of pathogens and their hosts. *Nature Genetics* 32:569–577.

Wright, S. 1943. Isolation by distance. *Genetics* 28:114–138.

———. 1982. Character change, speciation, and the higher taxa. *Evolution* 36:427–443.

Yeh, P., and T. Price. 2004. Adaptive phenotypic plasticity and the successful colonization of a novel environment. *American Naturalist* 164:531–542.

Yoder, J. B., and S. L. Nuismer. 2010. When does coevolution promote diversification? *American Naturalist* 176:802–817.

Yoshizawa, K., and K. P. Johnson. 2013. Changes in base composition bias of nuclear and mitochondrial genes in lice (Insecta: Psocodea). *Genetica* 141:491–499.

Young, N. D., and C. W. dePamphilis. 2005. Rate variation in parasitic plants:

correlated and uncorrelated patterns among plastid genes of different function. *BMC Evolutionary Biology* 5:16.

Yu, D., and H. Wilson. 2001. The competition-colonization trade-off is dead; long live the competition-colonization trade-off. *American Naturalist* 158:49–63.

Zamma, K. 2002a. Grooming site preferences determined by lice infection among Japanese macaques in Arashiyama. *Primates* 43:41–49.

———. 2002b. Leaf-grooming by a wild chimpanzee in Mahale. *Primates* 43:87–90.

———. 2006. A louse left on a leaf. *Pan Africa News* 13:8–10.

Zangerl, A. R., and M. R. Berenbaum. 2003. Phenotype matching in wild parsnip and parsnip webworms: causes and consequences. *Evolution* 57:806–815.

Zietara, M. S., and J. Lumme. 2002. Speciation by host switch and adaptive radiation in a fish parasite genus *Gyrodactylus* (Monogenea: Gyrodactylidae). *Evolution* 56: 2445–2458.

Zimmer, C., and D. Emlen. 2013. *Evolution: Making Sense of Life*. Roberts, Greenwood Village, CO.

Zinsser, H. 1935. *Rats, Lice and History*. Little, Brown, Boston.

Zohdy, S., A. D. Kemp, L. A. Durden, P. C. Wright, and J. Jernvall. 2012. Mapping the social network: tracking lice in a wild primate (*Microcebus rufus*) population to infer social contacts and vector potential. *BMC Ecology* 12:4.

INDEX

Britt, A. G., 43

Brockhurst, M. A., 13, 21, 212, 216, 221

Brodie, E. D., III, 10, 18

Broennimann, P., plate 7

Bronstein, J. L., 6

brood parasites
    cuckoos, 6, 130
    lice of, 130

Brooke, M. de L., 73, 127–28, 130

Brooks, D. R., 4, 157–58, 212–13

Brouqui, P., 40

Brown, C. R., 48

Brown, G. K., 46, 110

Brown, J. K. M., 211

Brown, N. S., 44, 68

Brown, S. P., 108

Brucker, R. M., 222

Brueelia
    apiastri, 49, 128
    deficiens, 181
    galapagensis, 146
    horizontal transmission, 128–29, 135,
        plate 3
    host specificity of, 34, 159
    phylogeny, 181–83
    population structure, 144–45
    as possible cause of feather holes, 52
    semiannulata, 144–45

Brunetti, O., 47

Bruyndonckx, N., 158

Bubalus bubalis, 74

Buchnera, 30

Buckling, A., 14, 216, 221, 223

Bucorvus leadbeateri, plate 11

Bull, C. M., 120

Burgess, I., 41, 68, 76

Burkett-Cadena, N. D., 85

Bursera, 217, 219

Burtt, E. H., 62

Bush, A. O., 126

Bush, S. E., 41, 62, 68, 74, 89, 92–94,
        95–98, 101–3, 108, 114–18, 126, 132,
        142, 146, 167, 181, 183, 194, 204–5,
        plate 12

Busvine, J. R., 40, 98–99

Buteo galapagoensis, 35, 81, 132, 143, plate
    15

butterflies, 4

Buxton, P. A., 129

Cacatua galerita, plate 12

Caldwell, R. M., 57

Calyptorhynchus funereus, plate 12

Cameron, S. C., 174

Camillia japonica, 12, plate 1

Campanulotes compar, 93, plate 10
    competition with wing lice, 114–18
    damage to feathers, 50
    dispersal, 135–37
    escape behavior, 89–90

Campbell, J. B., 42

Canestrari, D., 6

Cannon, S. M., 86–87

Carpodacus mexicanus, 76

Castoridae, 38

Centrocercus urophasianus, 53

Ceratosolen, 215

character displacement, 109

Charadriiformes, 31, plate 16

Charleston, M. A., 157

Chen, B. L., 113

chewing lice
    body size of, 88
    diet of, 28–29, 51, 81, plate 3
    diversity of, 28
    hosts of, 25

chicken, 35, 44, 68, 75, 113–14, 133

Chilt, N. B., 158

chimpanzee, 169–72, 174–75

chloroplasts as mutualists, 5

Christensen, N. O., 106

Ciconiiformes, plate 16

clade-limited colonization. *See* host
        switching: preferential

Clark, L., 76

Clark, M. A., 212

Clay, T., 25–26, 86, 107, 113, 133, 183–84, 197

Clayton, A. L., 31

parasitism, definition, 5. *See also* parasites

Parchman, T. L., 216

Parsons, P. A., 118

Parus major, 126

Passer domesticus, 63, 76, 133

Passeriformes, plate 16

Pastinaca sativa, 7, plate 1

Patagioenas
  fasciata, 93, 163
  phylogeny, 163

Paterson, A. M., 37, 159, 163, 167–68, 223

Paterson, S., 221

Patton, J. L., 143

Pavo cristatus, plate 6

pectinate claw, 74

Pediculus
  evolutionary rates, 172
  mjobergi, 47
  phylogeny, 169–72
  schaeffi, 169–72, 174

Pediculus humanus
  attachment, 87
  diet of, 28
  immune response to, 80
  mitochondria, 173–74
  origin of clothing, 173
  phylogeny, 169–72
  population structure, 143
  saliva, 104
  symbiont of, 28
  as vector, 39–40

Pelecaniformes, plate 16

Pelecanus erythrorhyncus, 48

Percy, D. M., 157

Pérez, T. M., 29

Perotti, M. A., 28, 35

Perrin, N., 13

Persson, L., 117

Pessoaiella, 200

Peterson, A. T., 72

Pfeffer, A., 82

Pfennig, D. W., 106

Phasianus colchicus, 129

phenotypic interface of coadaptation, 9, 191

Philopteridae, 23

Philopterus, 144–45

Phoca vitulina, 47

Phoeniculus purpureus, 73

phoresis, 130, 134–37, 152, 180, 182, 189, plate 3

Phthiraptera
  fossils of, 31, plate 3
  order, 27

phyletic tracking, 212. *See also* codiversification

phylogenetic interface, of codiversification, 18, 191, 213

phylogenetic tracking, 212. *See also* codiversification

Physconelloides, 158
  ceratoceps, 148–49
  eurysema, 93, 161
  phylogeny, 159, 161, 163, 187
  population structure, 147–49
  spenceri, 163
  zenaidurae, 23

Piagetiella peralis, 48

Picea mariana, 216

Piciformes, plate 16

Piculell, B., 11

pigeon. *See Columba livia*; Columbiformes

Pinus contorta, 14, 216

Pitohui, 62

Pittendrigh, B. R., 98

plants
  defensive compounds of, 7, 16
  as hosts, 4, 217
  pollination of, 7, 9
  seeds of, 12, 14–15

Plasmodium chabaudi, 109

Podiceps nigricollis, 110

Poiani, A., 56

Poinar, G. O., 16

Pollack, R. J., 104

feather molt, 61
  genetics of, 81–84
  immunological, 78–81
  inbreeding and, 83–84
resource tracking, 217
Reznick, D. N., 13
Rhopaloceras rudimentarius, 86
Rhynchophthirina
  hosts of, 27–29
  suborder, 28–29
Ribeiro, J. M. C., 104
Rickettsia prowazekii, 40
Ricklefs, R. E., 199, 221, 223
Ries, E., 30
Ritter, R. C., 55
Roberts, L. S., 107
Roberts, R. J., 41, 68
rock pigeon. *See Columba livia*
Rodgman, A., 77
Rodríguez-Gironés, M. A., 7
Rohde, K., 110
Ronald, N. C., 47
Ronquist, F., 161
Rose, K. D., 65–66
Rossini, C., 76
rostral groove, 88
Rothschild, M., 21, 46, 86, 98, 107, 111,
  128, 176
Rothschild, W., 7
Rózsa, L., 34–35, 41, 184
Rudolph, D., 118
Rueesch, S., 78
Rundell, R. J., 142, 155
Rundle, P. A., 39
Rust, R. W., 143
Ruxton, G. D., 97

Sadd, B. M., 11
Saemundssonia lari, plate 2
Sage, R. D., 222
Saino, N., 53, 80
Samuel, W. M., 58
Sandre, S.-L., 98

Sangster, N. C., 104
Sarconema eurycera, 44
Sari, E. H. R., 146
Sasaki-Fukatsu, K., 28
Sasal, P., 87
Saxena, A. K., 43
Sazima, I., 78
Scheiner, S. M., 205
Schluter, D., 106, 109, 118, 155, 198, 200,
  202, 216
Schmid-Hempel, P., 46, 56–57, 81, 104,
  142, 162, 211, 221
Schmidt, A. R., 16
Schulte, R. D., 221
Schwartz, C. C., 47
Sciuridae, 38
scratching, 73–74, 78, 197, 202
seals, 36, 47, 127
Seegar, W. S., 44
Seehausen, O., 198
Seger, J., 140
Segraves, K. A., 158, 210, 217
selection. *See* natural selection; sexual
  selection
Serranus cabrilla, plate 1
Servedio, M. R., 142
setae, 85–86
sex ratio in lice, 35
sexual selection, parasite mediated, 50,
  55–56, 84, plate 6
Shapiro, M. D., 100
Sheldon, B. C., 56
Sibly, R. M., 91
Simková, A., 110, 164
Simmons, K. E. L., 76–77
Simmons, R. B., 141
Simpson, G. G., 121, 198
Singer, M. S., 14, 217
Sinistrofulgar, 16, 18
Skirnisson, K., 47
Smith, C. I., 19
Smith, J. W., 14
Smith, V. S., 28, 32–33, 86, 159